Global Navigation Satellite Systems

Global Navigation Satellite Systems

Cody Rhodes

MURPHY & MOORE
www.murphy-moorepublishing.com

Published by Murphy & Moore Publishing
1 Rockefeller Plaza,
New York City, NY 10020, USA
www.murphy-moorepublishing.com

Global Navigation Satellite Systems
Cody Rhodes

International Standard Book Number: 978-1-63987-260-2 (Hardback)

Cataloging-in-Publication Data

Global navigation satellite systems / Cody Rhodes.
p. cm.
Includes bibliographical references and index.
ISBN 978-1-63987-260-2
1. Artificial satellites in navigation. 2. Global positioning system.
3. Mobile geographic information systems. I. Rhodes, Cody.
TL798.N3 G56 2022
629.045--dc23

Table of Contents

Preface

It is with great pleasure that I present this book. It has been carefully written after numerous discussions with my peers and other practitioners of the field. I would like to take this opportunity to thank my family and friends who have been extremely supporting at every step in my life.

A system which provides autonomous geospatial positioning by using satellites is referred to as a satellite navigation system. It helps small electronic receivers in determining their location including their latitude, altitude and longitude to a high precision. The location is determined by transmitting time signals along a line of sight by radio devices from satellites. A satellite navigation system which provides a global coverage is known as a global navigation satellite system (GNSS). The primary functions of these systems are to help in navigation, determining the exact position and tracking the position of any object. Some of the categories in which the global navigation satellite systems can be divided are core satellite navigation systems, Continental scale Ground Based Augmentation Systems (GBAS), Global Satellite Based Augmentation Systems (SBAS), Regional Satellite Navigation Systems, etc. This book outlines the processes and applications of global navigation satellite systems in detail. It elucidates the concepts and innovative models around prospective developments with respect to GNSS. Coherent flow of topics, student-friendly language and extensive use of examples make this book an invaluable source of knowledge.

The chapters below are organized to facilitate a comprehensive understanding of the subject:

Chapter - Introduction

Satellite navigation system is referred to a system that provides geo-spatial positioning through the use of satellites. Global navigation satellite system is a combination of satellites that helps in transmitting the positioning and timing data to GNSS receivers. This is an introductory chapter which will briefly introduce about global navigation satellite system.

Chapter - Global Navigation Satellite System Augmentation

The method that is used for the improvement of navigation system's attributes like accuracy, reliability and availability, through the use of external resources is called GNSS augmentation. It has two major systems namely local area and wide area augmentation system. This chapter closely examines the method of GNSS augmentation to provide an extensive understanding of the subject.

Chapter - Global Positioning System

Global Positioning System is a type of global navigation satellite system that gives geolocation and time information to a GPS receiver anywhere on Earth. It consists of a network of 24 orbiting satellites. This chapter has been carefully written to provide an easy understanding of all the aspects of global navigation system.

Chapter - Global Positioning System Based Software

There are many GPS based software that are used for navigation and tracking purposes. Some of the software are Google Earth, Google Maps, Google Latitude, GPS Logger, etc. The topics elaborated in this chapter will help in gaining a better perspective of these different GPS based software.

Chapter - Geolocation

Geolocation is the process of estimating and generating the geographical coordinates of the real-world geographic location of an object such as a radar source, mobile phone and Internet connected computer terminal. Indoor positioning system, Wi-Fi positioning system, geographic system, etc. are some of the systems used in it. This chapter discusses the subject of geolocation in detail.

Chapter - Applications

GNSS has a varied range of applications in transportation, precision agriculture, aerial photogrammetry, marine applications, geospatial information systems, rails, roads, etc. All these applications of GNSS have been carefully analyzed in this chapter.

Cody Rhodes

1

Introduction

Satellite navigation system is referred to a system that provides geo-spatial positioning through the use of satellites. Global navigation satellite system is a combination of satellites that helps in transmitting the positioning and timing data to GNSS receivers. This is an introductory chapter which will briefly introduce about global navigation satellite system.

Satellite Navigation

Satellite navigation is a system which makes use of artificial satellites for providing autonomous geospatial positioning. In other words, satellite navigation systems can generate positioning information. Satellite navigation systems can also provide information regarding local time to high precision and can aid in time synchronization.

With the help of transmitted time signals and electronic receivers, satellite navigation can determine a specific location with high precision. However for accurately locating the position, a minimum of four satellites would be required. Electronic receivers make use of signals for analyzing the current local time to high precision. Satellite navigation capable of global coverage is called a global navigation satellite system. The electronic receiver analyzes the time and the distance from each of the satellites to reach the user's receiver antenna. The fourth satellite in the navigation system helps in removing the distance ambiguity involved in the case of three satellites. In fact, the more satellites involved in a satellite navigation system, the higher precision the accuracy that can be achieved.

Satellite navigation systems are more accurate than other navigation options. Being a broadcast system based on radio signals, an unlimited number of people can use it an unlimited number of times, regardless of location. Satellite navigation systems aid in air and sea traffic as well.

GNSS

While some transmissions like TV and radio entertain us, others such as GPS provide invaluable information that we would quite literally be lost without. And yet despite the importance of GPS in our daily lives, we generally understand very little about how it works.

This document is designed to shed a little light on the subject of satellite navigation. It goes deeper than some explanations, but avoids the complicated things only system engineers need to know. By the end of it, you'll have strong understanding of how satellite navigation actually works, and more importantly what it can, and can't, do for you.

The term GPS is now so familiar to the general public that it has become synonymous with any satellite navigation system—but we need to be careful. The term GPS is actually the name of the system created by the American military, and stands for Global Positioning System. There are other systems too. The Russian military system for example is called GLONASS (Global Orbiting Navigation Satellite System), while China is developing a system called BeiDou-2.

When talking about satellite navigation systems, a term you will often encounter is GNSS, which stands for Global Navigation Satellite System. This is the generic name used to describe any global system of satellites that transmit signals for navigation purposes on Earth. The distinction of using terms like GPS or GLONASS is important if you're using satellite systems for measurement purposes because each system works in a slightly different way (although the end result appears to be the same).

As of 2017, the main GNSSs are:

- GPS: America's Global Position System.
- GLONASS: Russia's Global Orbiting Navigation Satellite System.
- BeiDou-2: A system being developed by China—not currently fully operational.
- Galileo: A system being developed by the EU—not currently fully operational.

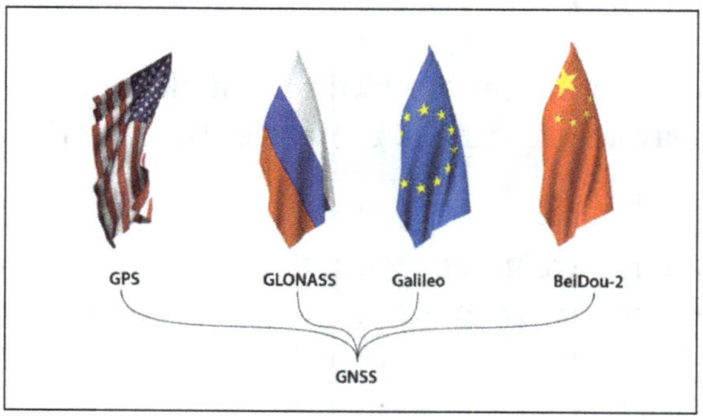

Global navigation satellite systems.

How does a GNSS Receiver Work?

There are two parts to most GNSS receivers; the antenna and the processing unit or receiver. The antenna is where the satellite signals are received, while the receiver makes sense of the information being received and turns it into measurements we understand, such as latitude and longitude.

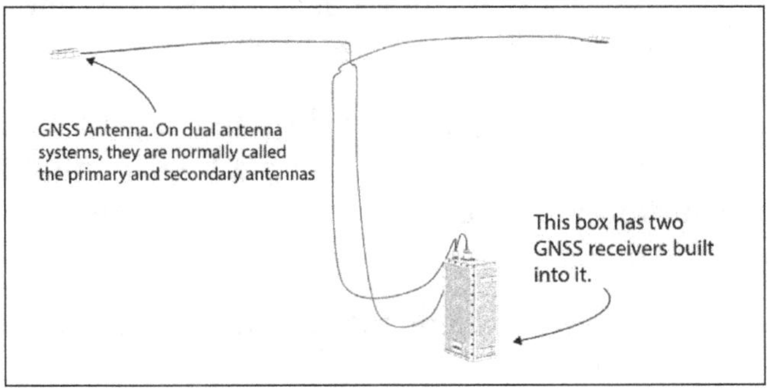

Although the GNSS receivers do all the work, the actual measurements they produce relate to the position of the antennas themselves. This is important to bear in mind, because the length of the antenna cables means a receiver can sometimes be quite a distance from the output position measurements. On sat navs and everyday GPS products, this doesn't matter as they are rarely capable of accuracy greater than several meters anyway.

It's important to recognize that the calculations about position, speed and altitude relate to the antenna itself, and not the receiver. To understand how a GNSS works, we need to break the GNSS into parts and understand a little bit about each one. As GPS is the system people are most familiar with we'll just look at that, and break it down into three parts:

- The space segment.
- The control segment.
- The user segment.

Space Segment

The space segment is concerned with the satellites in orbit. In 2015 the GPS constellation consists of 32 non-geostationary satellites in medium Earth orbit, although not all satellites are active. Each satellite orbits once every 11 hours, 58 minutes and 2 seconds at an average altitude of 20,200 km (that's an orbital radius of 26,571 km).

The GPS satellite constellation is arranged into six equally-spaced orbital planes, with no fewer than four satellites in each plane. This arrangement ensures a minimum of four satellites can be seen 15° above the horizon at almost any time, from any point on the planet—although in reality there are generally more.

While satellites vary in age and design, their principle of operation remains the same. Each one contains four highly accurate clocks with a fundamental frequency of 10.23 MHz, and they and constantly transmit two carrier waves in the L-Band that travel back to earth at the speed of light. These carrier waves are referred to as L1 and L2.

- The L1 carrier has a frequency of 1575.42 MHz (10.23 MHz × 154 = 1575.42 MHz).
- The L2 carrier has a frequency of 1227.60 MHz (10.23 MHz × 120 = 1227.60 MHz).

The carrier waves are important because they bring the information from the satellite back to earth, and it's that information that allows our receiver to work out where we are. But we'll come back to how this is done later.

Control Segment

The control segment refers to a number of ground stations situated around the globe (close to the equator) that are used to track, control and send information to each of the GPS satellites. This is an important role as it is vital the clocks in each satellite are synchronized—because the whole system relies on timing.

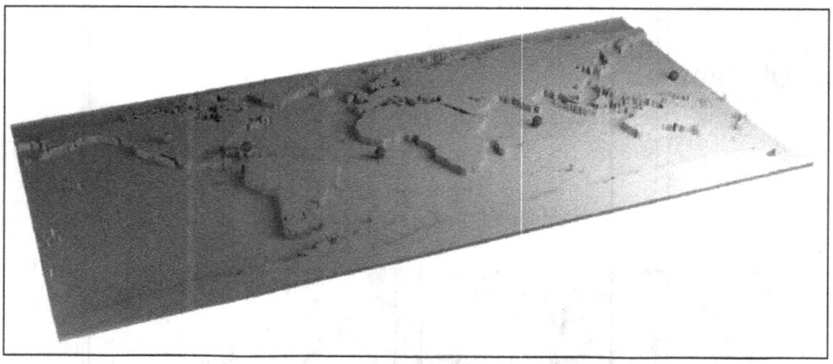

Four ground antennas (red) send comments from the master control station (yellow) to satellites in the GPS constellation.

The orbit information that is sent up to each satellite is vital too, because we need it in order to work out where the satellite was when the information was sent. All of this information is sent up to the satellites, then carried to your GPS receiver within the L1 carrier wave navigation message.

User Segment

The user segment is the part most people are interested in. This segment includes anyone or anything with a GPS receiver; sat navs, mobile phones, UAVs, law enforcement. So how does it work?

There is a constellation of satellites orbiting above our heads, sending a constant stream of information back to earth at the speed of light. Understanding how this helps pinpoint our location takes a bit of time, but it is based on a process called trilateration.

At no point does the GNSS receiver inside your sat nav or phone send any information up to the satellites. The receivers we use today are completely passive—they only receive information. When Europe's Galileo system is operational, its receivers will be slightly different because there will be an emergency function, which will send information when activated, but this won't apply in normal operation.

When you hear people talking about something being tracked by GPS, an armoured car for example, what is happening is this. The GNSS receiver on the vehicle is receiving signals from the satellites and working out where it is. Once it knows its position, it sends this information using some other system, a GSM data connection for example, back to some monitoring station.

GPS products are completely passive. They only receive signals, they do not transmit anything.

So, for the time being, GNSS receivers work by receiving signals sent from the relevant satellites in orbit. The signals that are used depend on the type of receiver. A GPS receiver can only make use of signals from the GPS satellites, while a GLONASS receiver can only use signals from GLONASS satellites. There is another kind of receiver that can actually take signals from both types of satellites though (GPS and GLONASS), to augment its measurements.

How does a GPS Receiver Work Out where I am?

The short and oversimplified answer to how a GPS receiver works out where you are is:

- It calculates how long it takes the signal from each satellite it can see to arrive.
- It multiplies that time by the speed of light to calculate the distance to each satellite.
- It then calculates its position relative to no fewer than three satellites using trilateration.

- Because the receiver knows the precise position of each satellite when the signal was sent, it can translate its own relative position into an Earth-based co-ordinate system.

Of course, this explanation causes an explosion of questions in most minds. How can a receiver calculate the travel time of the signal unless it knows when it was sent—and how could it possibly know that? What is trilateration? How does the receiver know where a satellite was when it sent the signal? And that's just the beginning.

GPS Signal

The hardest part of understanding GPS is the signal itself. Calculating distance based on time is quite straight forward, and even trilateration, isn't as tricky as it sounds when you add a few diagrams. But the GPS signal and how it works is pretty involved.

So what information is sent down to us? The unique signal transmitted from each satellite contains two codes and a message:

- C/A code (coarse/acquisition code).
- P code (precision code) [called Y code in its encrypted form].
- Navigation message.

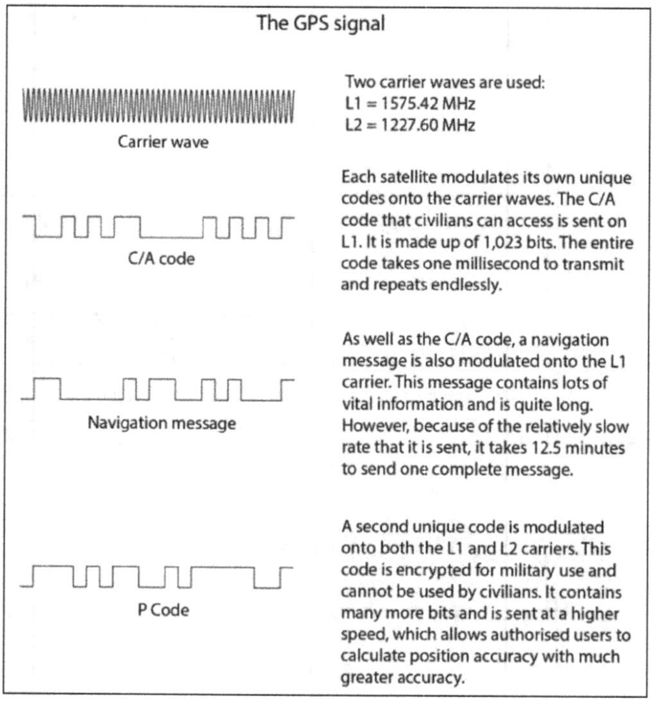

The P code is encrypted for military use, and so can be ignored. It's encrypted to stop spoofing and to control who has access to the system. Incidentally, once the P code has been encrypted, it's referred to as Y code.

As civilians, we're only interested in the C/A code and navigation message. The C/A code is actually a binary string made up of 1,023 bits. At first glance the sequence appears to be completely random—but it isn't. The sequence has been carefully composed so that if you wrote the sequence out on two pieces of tracing paper and overlaid them, there is only one position where they will match.

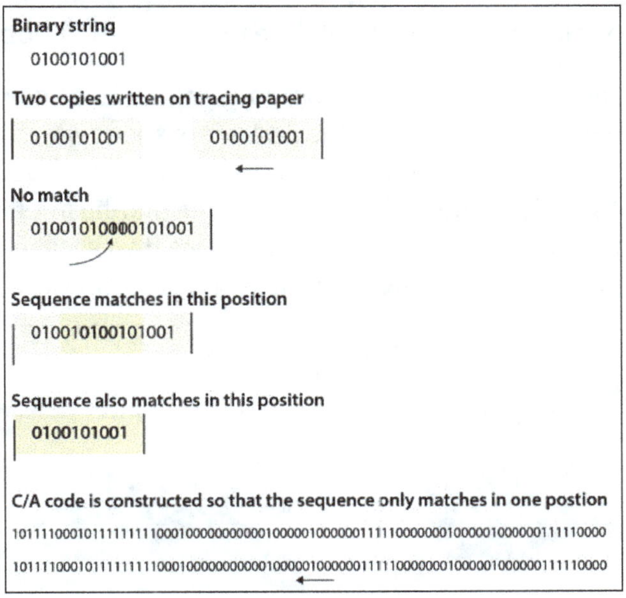

We'll look at why it's important the C/A code only matches in one position later when we look at how distance from the satellite is calculated. For now, all we need to know is that the C/A code each satellite transmits is unique to that satellite and that while it appears to be a random string of bits, it is not. The code actually follows a precise deterministic pattern, and so it is often called pseudo-random noise (PRN).

The navigation message is different from C/A code because it contains data. This includes information about clock corrections, satellite health, ephemeris (precise orbit) data, ionosphere model parameters and almanac (general information about all satellites in the constellation) data. Because of the amount of information contained in the navigation message, and the relatively slow rate that it's transmitted, it takes 12.5 minutes to send the whole message.

The final part of the GPS signal we need to look at is the carrier itself. As we've already said, each satellite transmits two frequencies—L1 at 1575.42 MHz and L2 at 1227.60 MHz. All three elements (the C/A code, Y code and navigation message) are modulated onto the L1 carrier, while only the Y code is modulated onto the L2 carrier. However, while civilian users can't demodulate the Y code from the L2 carrier, we can make use of the L2 carrier wave itself.

The C/A code is modulated onto the carrier wave at 1.023 Mb/s. The Navigation message is modulated onto the carrier at 50 b/s.

Finding Satellites

The first thing a receiver needs to do when it's powered up is look for satellites. How it does this depends on long it has been since the receiver was last locked on to a GPS satellite. That's because as we've already seen, each satellite in the GPS constellation transmits a navigation message, and part of that message is an almanac that contains rough information on the orbit of every satellite in the constellation.

Almanac data is useable for about 6 months, so using its last known position and the current time, the receiver can make an educated guess about which satellites should be in view. It tries this because the alternative is to blindly search for all satellites—and that can take some time. In fact, that's why a receiver that hasn't been powered up for some time, or has been shipped around the world since it was last powered on, can take a great deal of time to acquire GPS lock—because it has to blindly search.

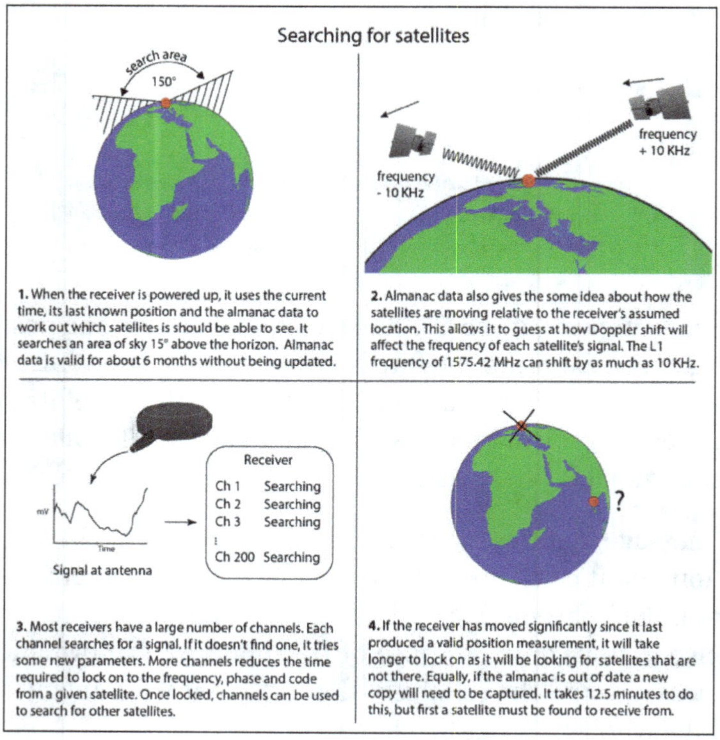

1. When the receiver is powered up, it uses the current time, its last known position and the almanac data to work out which satellites is should be able to see. It searches an area of sky 15° above the horizon. Almanac data is valid for about 6 months without being updated.

2. Almanac data also gives the some idea about how the satellites are moving relative to the receiver's assumed location. This allows it to guess at how Doppler shift will affect the frequency of each satellite's signal. The L1 frequency of 1575.42 MHz can shift by as much as 10 KHz.

3. Most receivers have a large number of channels. Each channel searches for a signal. If it doesn't find one, it tries some new parameters. More channels reduces the time required to lock on to the frequency, phase and code from a given satellite. Once locked, channels can be used to search for other satellites.

4. If the receiver has moved significantly since it last produced a valid position measurement, it will take longer to lock on as it will be looking for satellites that are not there. Equally, if the almanac is out of date a new copy will need to be captured. It takes 12.5 minutes to do this, but first a satellite must be found to receive from.

On the assumption that our receiver has valid almanac data, and hasn't moved much since it was last powered up, it will have a reasonable idea of which satellites it should be able to see at the current time and how the signal frequency will be affected by Doppler shift. It can now start to look for the signal from the satellites it expects to see in the signal the antenna is picking up.

The receiver begins by conducting a two-dimensional search for the signal's frequency and C/A code phase. This is not easy, but as end-users, we don't need to worry too much about this. Once the receiver thinks it has found the signal from a satellite it confirms this, then locks onto it with a PLL (Phase Locked Loop) and a CLL (Code Locked Loop). It then searches for bit synchronisation before demodulating the data.

Working out the Range to a Satellite

As we've already seen, each satellite in the GPS constellation transmits a unique C/A code, which appears to be random noise, but is in fact a deterministic sequence. Because the sequence is deterministic, a GPS receiver has the ability to generate exactly the same sequence of code as the satellite itself. By generating the same pattern internally, the receiver looks for the pattern being transmitted by the satellite, then works out how much of a delay there is compared to its own pattern.

It can do this because of the accurate clock synchronisation. For instance, imagine you and a friend are stood in a very large field. You synchronise your watches and your friend agrees to fire a starting pistol at exactly midday. Your friend then walks off to the other end of the field. At precisely midday, they fire the pistol, but according to your watch, you don't hear the bang until four seconds after mid-day because the sound had to travel across the field. If we assume the speed of sound is 340 m/s, we can multiply that by the time taken work out that your friend is 1,360 metres away.

The receiver and GPS satellite work in a similar way. Because the receiver is able to accurately synchronise its clock with those of the satellites, and because it knows what the time was when the speed by the time taken to work out how far away the satellite is.

There are other factors it needs to take into account of course, but that is the underlying principal.

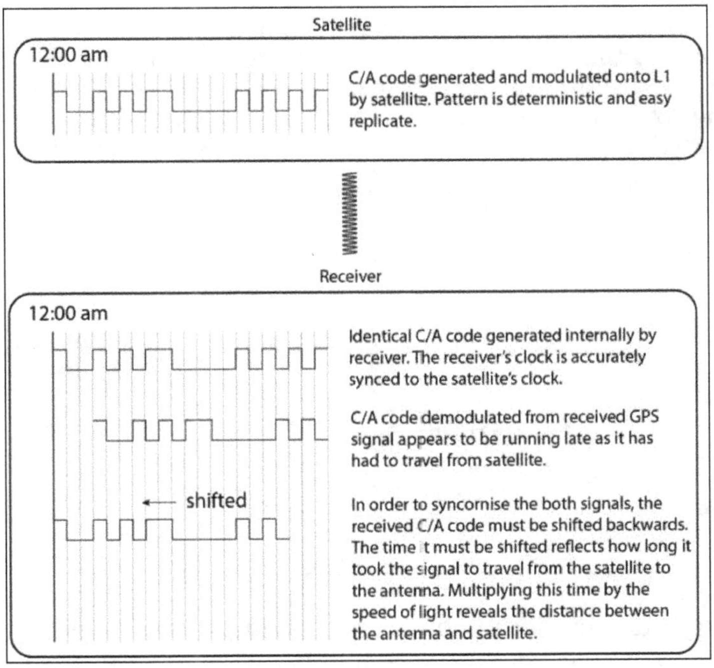

How can Distance Measurements Help Me Work Out where I Am?

Continuing with the example of you and some friends in a field, image you are stood in a large square field. Each friend has a way of making a different noise (starter pistol, cymbal, shouting, etc.). You synchronise watches, and one friend goes and stands in each corner of the field. Starting at mid-day, they agree to start making a noise once every 10 seconds.

Now suppose when they start to make noise that two of the sounds arrive simultaneously, shortly followed by the other two. You then move a little in the direction of delayed sounds. Ten seconds later another burst of noise arrives, but this time the delay for the second pair is shorter. If you keep moving you will get to a point where all four noises arrive simultaneously. As you know each person is stood in the corner of the field, you can also work out that you must be stood exactly at the centre.

In terms of GPS navigation, the satellites may not be nicely located in the corners of the sky, but because of the ephemeris data, we can work out their precise location relative to a known co-ordinate system when the signal was sent, and using the C/A code, we can also work out how far we are from that know location. What happens next is a process called trilateration.

Trilateration is a bit like triangulation. With triangulation, you identify a specific point by saying it is at angle of a from point 1 and angle of b from point 2. Lines drawn at those specified angles from each point will cross, and the point at which they cross is the location of our new point.

Triangulation

Triangulation projects lines of unknown length along known angles to find a point. As long as there is more than one reference point we can identify the location of a new point. So, if we know point *x* is located at angle of 45° from point 1, and at an -45° from known point 2, the point at which those projected lines intersect must be the location of point *x*.

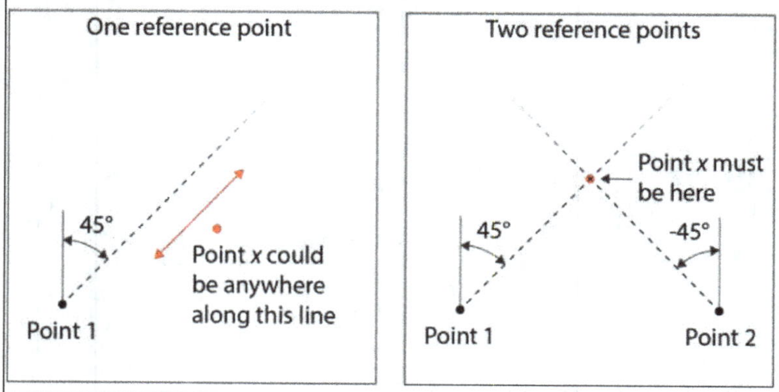

Triangulation projects lines of unknown length along known angles to find a point. As long as there is more than one reference point we can identify the location of a new point. So, if we know point x is located at the angle of 45° from point 1, and at an -45° from known point 2, the point at which those projected lines intersect must be the location of point x.

Trilateration works in a similar way but uses distance rather than angles to find a point. The other major difference is that with trilateration, you need a minimum of three reference points rather than two to narrow the search down to one location.

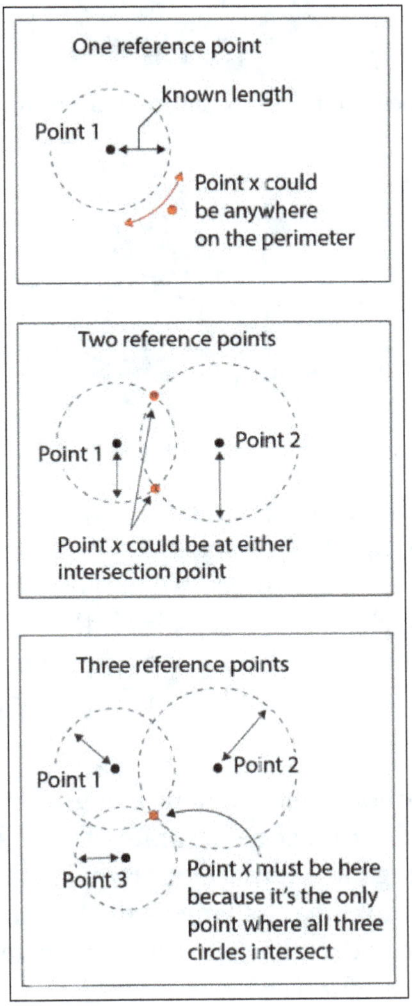

Trilateration uses lines of known length but unknown angles (circles in other words) to find a point. As long as there are more than two reference points we can identify the location of a new point. So, if we know point x is 1 metre from point 1, 1.5 metres from point 2 and 0.75 metres from point 3, the point at which those circles intersect must be the location of point x.

GPS uses the same technique but has to approach things slightly differently. The circles we've been looking at are two-dimensional. In real life, things are far more three-dimensional, and that means our intersecting circles become intersecting spheres.

When two spheres intersect, you don't end up with two points of intersection, you end up with a ring of intersection (imagine two bubbles joined together). If a third bubble joins in, it creates two points where all three intersection rings meet.

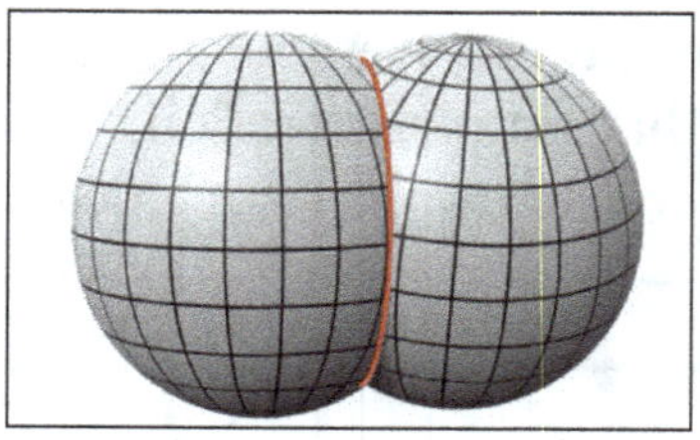

When two spheres intersect, the intersection creates a circle (red).

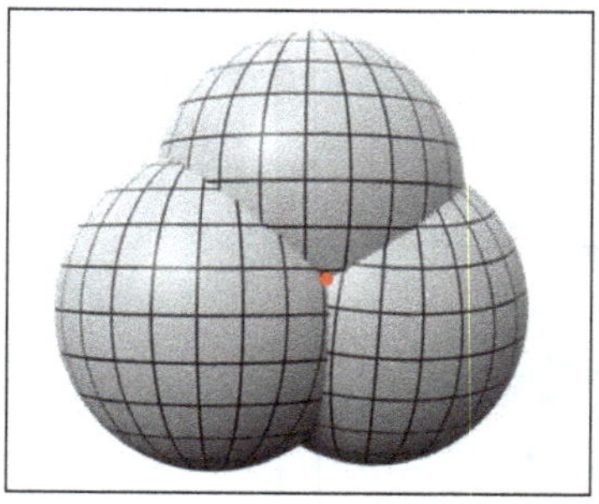

When three spheres intersect, there are only two points common to all three spheres (one each side).

At this point, you might recall hearing that GPS only needs three satellites in order to generate a position measurement. This statement is both right and wrong. It's clear from the image above that in order to use trilateration to arrive at a single point, a fourth sphere is required (to show us which point is the correct one). However, as GPS uses an Earth centred, Earth fixed coordinate system, one of the points can be ruled out immediately as it would be well outside Earth's atmosphere. The other point must, therefore, be the correct one.

So GPS can generate a position measurement using only three satellites, but in order to do that, its internal clock must be accurate otherwise it can't calculate the distance properly. And correcting the internal clock requires four satellites! That's why the statement was both right and wrong. As long as the system has used four satellites to correct its internal clock, it can drop back to only three satellites and still make a position estimate—it just can't do it forever.

What is SPS?

The term SPS stands for Standard Positioning Service, and describes GPS position measurements that are based only on the C/A code. The distinction is needed because not all GPS position measurements are made in the same way, and the method used to calculate the measurement affects the accuracy.

SPS provides the lowest accuracy GPS position measurements, normally in the region of 3–10 metres. To make SPS measurements the GPS receiver locks onto four or more satellites, and then uses the C/A code to estimate the distance to each satellite. These estimates are called pseudo-range measurements.

Using SPS, the accuracy of the measurement comes from the receiver's ability to correctly align its internally generated C/A code with that received at the antenna. Although even if it were able to do this perfectly, the accuracy is still limited by the design of the system.

In most cases, the best alignment accuracy we can hope to for is about 1% of the width of a bit. As bits are transmitted at 1,023 per millisecond, and light travels about 300,000 metres per millisecond, our 1% alignment gives us an accuracy of roughly three metres. However, some L1 receivers can achieve alignment to roughly 0.6%, giving them an accuracy of 1.8 metres.

While that level of accuracy is adequate for sat navs, greater accuracy is required in many other applications, so other measurement techniques such as DGPS (Differential GPS) and Real-time Kinematics (RTK) are often required.

What are Differential Corrections or DGPS?

Most of the error in an SPS measurement comes from the fact the receiver has no way of actually knowing how the speed of the satellite signal has been affected as it passed through the ionosphere. It gets a small clue as there are some estimates included in the navigation message, which allow it to make a guess at the likely effects, but these are only estimates. In practice it means the speed by which the delay is multiplied is not quite correct—that, in turn, means the calculated pseudo-range is also incorrect.

The only way to allow for this is to measure, rather than estimate, how the signal speed is being affected as it makes its way down through the atmosphere. This is where differential GPS come in. DGPS can be achieved in a number of ways, but they essentially use the same technique.

A separate GPS receiver (called a base-station) is placed at a pre-surveyed point and turned on. Because the location of the base station is already known with great accuracy, the base station is able to compare the position measurement generated by its own GPS receiver with the known co-ordinates. Any difference means they signal from one of more satellites is being delayed. All the system has to do then is work out how much correction should be applied to each satellite in order to correct the GPS position measurement.

Once the base station knows how the signal from each satellite needs to be corrected, it can share this information with any other GPS receivers (often called rovers) in the area. This is often achieved using radio modems. The roving receiver checks if corrections are available for the satellites it is seeing and applies them where applicable—improving the accuracy of its pseudo-range measurements even though it is not at a known location.

There are also other ways to achieve DGPS. Rather than using your own local base-station (that you set-up on your own site), it is possible to receive differential corrections via a web-based service and data connection. In this case, your receiver (the rover) works out where it is in the normal way. It then transmits its location to a central server that works out where the nearest reference station is to you, before sending back the differential corrections from that location. The beauty of systems such as these is that you could drive from one end of a country to the other and receive valid corrections for the entire journey.

Another method is to use satellite-based correction services called SBAS (Satellite Based Augmentation System). These work in a slightly different way. Ground stations on earth at known locations work out the differential corrections in different geographic locations. These corrections are then uploaded to the satellite where they can be broadcast down to any receiver that is enabled to receive them. However, as well as needing a receiver with the correct hardware, you often have to pay a subscription fee in order to unlock the corrections.

Using SBAS services improves the accuracy of receivers, but not to the same degree as a local base station. When using DGPS, the accuracy of position measurements often improves to tens of centimetres, say around 40 cm. DGPS that is achieved through SABS is often sub-metre accurate, somewhere in the range of 50–60 cm.

What is RTK?

RTK stands for Real-time Kinematic and is another technique that improves the accuracy of GPS position measurements—however, it's one of the hardest to understand and the most intensive to implement.

We've already seen that SPS is the first position fix that GPS systems achieve, and they do this using the C/A code. If differential corrections are available, the SPS accuracy can be improved as timing delays on the satellite signals can be removed. Using DPGS position accuracy of around 40 cm is fairly normal. RTK is the next step on from DGPS, and there are two versions of it; RTK float, which achieves decimetre level accuracy and RTK integer, which achieves centimetre level accuracy.

To understand how either version works it's necessary to understand two things. The first is that RTK is bending the rules of what we should be able to achieve using a civilian GPS receiver. It is not a technology that was designed into the system to help us. Instead, it's a method that was figured out by GPS manufacturers to achieve greater accuracy than was intended using the C/A code. The second thing to understand is that it is based on the carrier waves themselves, not on the C/A code or navigation message they carry.

So how does it work? We've already seen that SPS is based on pseudo-range measurements calculated using the C/A code. The receiver does this by synchronising its clock with the GPS satellites, then generating its own version of the C/A code for each satellite it can see. If it has to delay its own copy of the C/A code by, say seven milliseconds, in order to match the C/A code being received at the antenna, then it knows the travel time of the signal from that satellite is seven milliseconds and it can then work out how far away the satellite is.

The ultimate aim of RTK is to establish how many carrier waves there are between the antenna and the satellite. The reason for this is simple. We've already seen that each satellite broadcasts a unique C/A code made up of 1,023 bits. The code is sent at a rate of 1.023 Mb/s, which means one bit is sent about every microsecond. In one microsecond the radio signal from the satellite covers a distance of about 300 metres.

The carrier wave that the C/A code is modulated onto is at a much higher frequency however—1575.42 MHz. This means a single wave covers about 19 cm. If we could work out how many full waves there are were between the satellite and the antenna, then it would be possible to calculate the distance much more accurately. In fact, if we know how many full waves there are, and it's possible to measure partial waves (the phase angle) too, then we can be very precise.

RTK Integer and RTK Float can be considered as two separate (but related) algorithms running concurrently within a system. If RTK Integer has a valid solution, the system switches to it (however the RTK Float algorithm continues running in the background). If RTK Integer subsequently becomes invalid, the system reverts back to RTK Float, which does not need to start again as long as the carrier phase lock is not lost.

So how are the algorithms different? RTK Float aims to identify your probable location (improving on the current DGPS accuracy) using statistical methods. It requires a minimum of four common satellites with the base station, and (in lay-terms), looks for a point in a circle around the current position measurement about which the satellites are revolving. Unlike RTK Integer, the Float algorithm never attempts to solve the ambiguity problem—it only tries to identify the most probable position within a circle drawn around the current position estimate. The accuracy for RTK Float starts around 40 cm, but increases to 20 cm at best.

RTK Integer, on the other hand, aims to solve the ambiguity problem and is not used until it has done so. It requires five common satellites, and when a valid solution is found the system knows there are n carrier waves plus any partial wave between it and the satellite. It can align its measurements to about 0.6% of the 19 cm wavelength, giving an accuracy of about 1 cm.

Doesn't DGPS or RTK Improve the Accuracy of Velocity Measurements?

It's a common misconception that GPS velocity measurements are made by measuring the time taken to move between two consecutive position measurements. This is not how velocity is measured—and is also the reason why position accuracy has no effect on velocity accuracy.

To work out velocity, GPS receivers actually use Doppler shift. So they look at the frequency shift in the carrier wave from each satellite and calculate a velocity vector using that information. So it doesn't matter if your receiver is in SPS mode of RTK Integer, the velocity measurements will be equally accurate.

What are the Limitations of GNSS?

GNSS systems, on the whole, are pretty awesome. They're easy to use, don't drift and can achieve high levels of accuracy. However, they're not perfect. For starters, to reach their full potential they need a clear and uninterrupted view of the sky. That's great if you're working in the middle of a field, but if you're trying to survey city streets, or working under bridges and in tunnels, at best you get reduced accuracy, and at worst you get no measurements at all.

Also, while a GPS receiver is great at capturing position and velocity measurements it's not much good if you're also interested in roll, pitch and yaw. A dual-antenna system helps solve some of those problems, but still doesn't provide a complete picture. And for extremely dynamic manoeuvres, there is simply not the resolution required.

GNSS Reflectometry

By GNSS positioning a multipath signal is often considered an undesirable phenomenon that needs to be suppressed. A reflected GNSS signal (also known as multipath- ore scattered signal) is regarded as an error source that lower the positioning accuracy. But these scattered signals can be used in many remote sensing applications.

The term GNSS Reflectometry (GNSS-R) comprises various remote sensing techniques. GNSS-R aims to retrieve information about the Earth surface by analysing the signals emitted by GNSS-satellites and captured the reflected signal by an elevated platform. The system has the characteristics of a bi-static radar and scatterometre, at L-band (microwave). In this process, the GNSS L-band satellite acts as the transmitter and an airplane or low earth orbit (LEO) satellite, as the receiving platforms. For altimetry applications, a GNSS-R receiver can also be placed on the land.

The advantages of GNSS-R remote sensing:

- No additional transmitter (over satellite scatterometry and radar altimetry).
- Highly precise, continuous: more than one hundred transmitting GNSS satellites in the near future, including GPS, Galileo, GLONASS, and Beidou/Compass.
- All-weather day and night and near-real-time remote sensing tool.

- Use of spread-spectrum communication technology to enable the receiver to receive weak signals.
- Wide range of uses for such things as sea-wind retrieval, seawater salinity detection, ice-layer density measurement, humidity measurement of land, and detection of moving targets.

GNSS Applications

GNSS technology has been adopted by the consumer market, in an ever-increasing range of products.

GNSS receivers are now routinely integrated into smartphones, to support applications that display maps showing the location of and best route to stores and restaurants.

Portable navigation devices give drivers directions on road or off.

Portable navigation device.

Geocaching is an outdoor activity in which participants use a GNSS receiver to hide and seek containers (called "geocaches" or "caches") around the world.

Currently, most GNSS consumer products are based on GPS, but this will change as more GNSS constellations are implemented.

Transportation

In rail transportation, GNSS is used in conjunction with other technologies, to track the location of locomotives and rail cars, maintenance vehicles and wayside equipment, for display at central monitoring consoles. Knowing the precise location of rail equipment reduces accidents, delays, and operating costs, enhancing safety, track capacity, and customer service.

In aviation, GNSS is being used for aircraft navigation from departure, en route, to landing. GNSS facilitates aircraft navigation in remote areas that are not well served by ground-based navigation aids, and it is a significant component of collision avoidance systems, and of systems used to improve approaches to airport runways.

In marine transportation, GNSS is being used to accurately determine the position of ships when they are in open sea and also when they are maneuvering in congested ports. GNSS is incorporated into underwater surveying, buoy positioning, navigation hazard location, dredging, and mapping.

In surface transportation, vehicle location and in-vehicle navigation systems are now being used throughout the world. Many vehicles are equipped with navigation displays that superimpose vehicle location and status on maps. GNSS is used in systems that track and forecast the movement of freight and monitor road networks, improving efficiency and enhancing driver safety.

Port Automation

Using GNSS, shipping hubs can improve their operating efficiency by tracking the movement and placement of containers about their yards.

Gantry crane in the Port of Vancouver.

Gantry cranes are used in ports throughout the world to lift shipping containers. These cranes are large and sometimes difficult to steer accurately in a crowded shipping dock. Many cranes are equipped with GNSS-based steering devices that determine the crane's position and keep it travelling in the desired path, improving accuracy and productivity as well as the safety of operators and workers on the ground. A key benefit is the quick movement of containers about the port, which reduces food spoilage and gets toys delivered on time.

Driver Testing in China

The automotive market is booming in China. In 2001, there were only 5 million cars driving the roads in China. By the end of 2012, the number of cars had reached 120 million. This huge increase in cars also means a huge increase in new drivers. To manage the testing of all these new drivers, the Chinese government has instituted an automated test as part of the driver's license test. In the automated test, the new drivers are required to perform a series of standard driving tasks, such as parking, turning, stopping, etc., on a surveyed course at a licensing facility. To automate the test, the licensing facilities use a GNSS system installed in the car to monitor and evaluate the driver. The

GNSS system installed on the car uses dual antennas so that the GNSS system can not only monitor the car's position and speed, but also detect the direction the car is going.

Parking Automation

In the Canadian city of Calgary, paying for on street parking has become automated. Customers pay for parking at street side terminals or using their smartphone and monitoring of the parked vehicles is done from a vehicle equipped with cameras and GNSS receiver.

When a customer pays for parking, they enter the license plate of their vehicle, a code that identifies the parking area and the amount of parking time required. This information is sent to a database. As the monitoring vehicle drives along the street, the vehicle cameras capture the license plates of the parked cars. The license plate number, along with the time and position provided by the GNSS receiver, is compared to the database of paid parking. If a vehicle is not found in the database, the photograph is sent to a Calgary Parking Authority employee so they can determine if there is a just cause (e.g. people are just getting out of car), there is a mistake in identifying the license plate (e.g. a D is mistaken for an O) or if it is a parking violation.

Due to the tight urban corridors in downtown Calgary, the location reported by the GNSS only system on the monitoring vehicles was misplacing 6-7% of the vehicles (approximately 1,400 vehicles) per day and the vehicles could be misplaced by up to 600 metres. This misplacement caused many hours of extra work each day for Calgary Parking Authority employees, because they have to manually correct the vehicles position before they can determine if there was a parking violation.

By switching from a GNSS only system to a GNSS+INS system, the monitoring vehicle was able to overcome the GNSS challenges in downtown Calgary and provide a much more reliable position. The GNSS+INS system reduced the number of misplaced cars to 1% (less than 300) saving the Calgary Parking Authority enough work hours to pay for GNSS+INS system in all of their monitoring vehicles in less than two years.

Machine Control

GNSS technology is being integrated into equipment such as bulldozers, excavators, graders, pavers and farm machinery to enhance productivity in the real-time operation of this equipment, and to provide situational awareness information to the equipment operator. The adoption of GNSS-based machine control is similar in its impact to the earlier adoption of hydraulics technology in machinery, which has had a profound effect on productivity and reliability.

Some of the benefits of GNSS-based machine control are summarized below:

- Efficiency: By helping the equipment operator get to the desired grade more quickly, GNSS helps speed up the work, reducing capital and operating costs.

- Accuracy: The precision achievable by GNSS-based solutions minimizes the need to stop work while a survey crew measures the grade.
- Job Management: Managers and contractors have access to accurate information about the jobsite, and the information can be viewed remotely.
- Data Management: Users can print out status reports, save important data and transfer files to head office.
- Theft Detection: GNSS allows users to define a "virtual fence" about their equipment and property, for the purpose of automatically raising an alarm when equipment is removed, then providing equipment tracking information to the authorities.

Precision Agriculture

In precision agriculture, GNSS-based applications are used to support farm planning, field mapping, soil sampling, tractor guidance, and crop assessment. More precise application of fertilizers, pesticides and herbicides reduces cost and environmental impact. GNSS applications can automatically guide farm implements along the contours of the earth in a manner that controls erosion and maximizes the effectiveness of irrigation systems. Farm machinery can be operated at higher speeds, day and night, with increased accuracy. This increased accuracy saves time and fuel, and maximizes the efficiency of the operation. Operator safety is also increased by greatly reducing fatigue.

Precisely aligned crop rows guided by GNSS-guided machinery.

Construction

GNSS information can be used to position the cutting edge of a blade on a bulldozer or grader, for example, or a bucket (excavator), and to compare this position against a 3D digital design to compute cut/fill amounts. "Indicate systems" provide the operator with visual cut/fill information, via a display or light bar, and the operator manually moves the machine's blade or bucket to get to grade. Automatic systems for bulldozers/graders use the cut/fill information to drive the hydraulic controls of the machine to automatically move the machine's blade to grade. Use of 3D machine control dramatically reduces the number of survey stakes required on a job site, reducing time and costs.

Productivity studies have repeatedly shown that the use of 3D machine control results in work being completed faster, more accurately and with significantly less rework than conventional construction methods.

Surface Mining

GNSS information is being used to efficiently manage the mining of an ore body and the movement of waste material. GNSS equipment installed on shovels and haul trucks provides position information to a computer-controlled dispatch system to optimally route haul trucks to and from each shovel. Position information is also used to track each bucket of material extracted by the shovel, to ensure that it is routed to the appropriate location in the mine (crusher, waste dump, leach pad). Position information is used by blast hole drills to improve fracturization of the rock material and control the depth of each hole that is drilled, to keep the benches level. Multi-constellation GNSS is particularly advantageous in a surface mining environment due to the obstructions caused by the mine's walls. More satellites mean more signal availability.

Automated Blast Hole Drilling

Automated drills are used in surface mines to increase safety and productivity. A single operator, located in the safer control room, can operate and monitor up to five automated drills.

The blast holes drilled by the automated drills must be very precise both horizontally and vertically. The position of the holes (horizontal accuracy) is critical in controlling rock fragmentation. Rock fragments that are too large or too fine can increase wear on the rock crushers used to process the material. Hole depth (vertical accuracy) is important for creating a flat bench.

Automatically guided trenching machine.

Three GNSS technologies are used on the automated drills, RTK, heading and multi-constellation. RTK provides the precise positioning needed to accurately locate the blast holes. Heading provides the alignment of the drill to ensure the holes are drilled perpendicular. Multi-constellation receivers compensate for signal blockages common in the high wall environment typical of surface mines.

Survey

GNSS-based surveying reduces the amount of equipment and labour required to determine the position of points on the surface of the Earth, when compared with previous surveying techniques. Using GNSS, it is possible for a single surveyor to accomplish in one day what might have taken a survey crew of three people a week to complete.

GNSS-based surveying equipement.

Determining a new survey position once required measuring distances and bearings from an existing (known) survey point to the new point. This required measurements using theodolites to measure angular differences and metal "chains" (long heavy tape measures), pulled taught to minimize sag and accurately measure distances. If the new and existing survey points were separated by a large distance, the process would involve multiple setups of the theodolite, then multiple angular and distance measurements.

Using GNSS, surveyors can now set up a DGNSS or RTK base station over an existing survey point and a DGNSS or RTK rover over the new point, then record the position measurement at the rover. This simplification shows why the surveying industry was one of the early civilian adopters of GNSS technology.

Seismic Survey Sensors

In a seismic survey, sound waves are sent from a source (explosives or a thumper truck) through the ground to an array of sensors (geophones). Knowing the exact location and orientation of the geophones is critical for a successful survey.

Using traditional geophones, the placement of the geophones was a two-step process. First a team surveys the area and places markers for each geophone. Later, a second team places the geophones precisely on the marked locations and then orients the geophones using a direction measuring device, such as a compass.

Using GNSS enabled geophones, eliminates the need to survey the area first. The GNSS enabled geophones have a GNSS receiver and dual antennas integrated in the geophone. The receiver and dual antennas allow the geophone to not only determine its exact location, but also its orientation.

Aerial Photogrammetry

Aerial photogrammetry refers to the recording of images of the ground photographs, for example, from an elevated position, such as an aircraft. Systems of this type are now more generally referred to as "remote sensing," since the images can be taken from aircraft or from satellites.

In the past, images would have to be manually corrected for orientation, perspective and the height of the camera and location, and manually "stitched" together. This manual process would be based on the accurate alignment of known points in adjacent pictures.

By integrating the camera with GNSS and INS, it is now possible to automate the process, in real time or post-mission, to "transfer" the location accuracy of the aircraft, determined from GNSS, to the image.

Aerial image of Niagara Falls.

Aerial photographs are used in online map systems such as Google Earth. Many of us have found our houses, and perhaps even our cars, through these applications.

GNSS technology has also been integrated with LiDAR (Light Detection and Ranging), an optical remote sensing technology used to measure range to distant targets. It is possible to image a feature or object down to the wavelength, which at LiDAR frequencies is less than a millionth of a metre.

Mapping Wildfires

To battle wildfires, fire fighters need to know the locations of the fires and any hotspots. Using an airplane equipped with an infrared imaging sensor and a GNSS+INS system, the locations of the fires and hotspots can be projected on topographical or 3D terrain maps.

Ground Mapping

Products have been developed that take 360-degree panoramic photographs to support the presentation of geometrically correct images on a computer screen. These images are continuous and precisely positioned. GNSS and IMU data is recorded before the panoramic photographs are taken. Position and attitude data is programmed into the cameras, allowing onscreen determination of positions of objects in the photos, or measurements between objects.

Infrastructure Visualization

Using a LiDAR system combined with a GNSS+INS system, a user can capture comprehensive visual information of key infrastructure, such as oil and gas pipelines. This visual information provides the state, location and positioning of the infrastructure and its related assets. It also assists in planning for maintenance and modifications.

Geospatial Information Systems

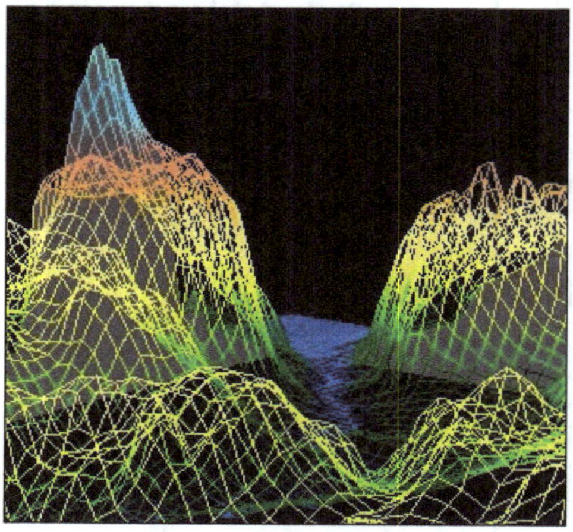

A geospatial information system (GIS) captures, stores, analyzes, manages, and presents data that is linked to location. The data may consist of, for example, environmental or resource data. GIS is also used to map attributes for insurance companies, municipal planning, utility companies, and others. The positions associated with the data can be provided from a GNSS receiver. GIS applications can generate detailed contour maps from the data and present these maps in a digital form.

Time Applications

As we mentioned time accuracy is critical for GNSS position determination. This is why GNSS satellites are equipped with atomic clocks, accurate to nanoseconds. As part of the position determining process, the local time of GNSS receivers becomes synchronized with the very accurate satellite time. This time information, by itself, has many applications, including the synchronization of communication systems, electrical power grids, and financial networks. GNSS-derived time works well for any application where precise timing is needed by devices that are dispersed over a wide area.

Seismic monitors that are synchronized with GNSS satellite clocks can be used to determine the epicenter of an earthquake by triangulation based on the exact time the earthquake was detected by each monitor.

Marine Applications

Marine applications of GNSS.

In addition to dramatically improving marine navigation, GNSS is also being applied to a broad range of marine applications, such as oilrig positioning, underwater cable and pipeline installation and inspection, rescue and recovery, and the dredging of ports and waterways.

GNSS-equipped Sonobuoys

An interesting application of GNSS is the use of GNSS-equipped sonobuoys in underwater sonar systems.

Sonobuoys are dropped from aircraft over an area of interest, but are left to float autonomously. The sonobuoys detect approaching ships and other hazards in the water by transmitting sound waves through the water, detecting reflections from vessels and objects, and determining the time it takes for the "echo" to be received. The data comes up to the sonobuoy's float then is transmitted, with GNSS positioning data, over a radio link to a survey ship. The survey ship collects and analyzes sonar data from a larger number of sonobuoys then determines and displays the location of ships and objects in the area of interest.

Seafloor Mapping

Knowing the depth of the seafloor in ports and navigation channels is critical to safe marine navigation. Maps of the ports and channels are created using bathymetric sonar systems.

Bathymetric sonar systems, mounted in a marine vessel, bounce sound waves off the seafloor to determine the depth of the water. Using these depth measurements, a map of the seafloor is created. For the seafloor maps to be accurate, the exact location of the vessel on the surface of the water must be known. A GNSS+INS system integrated with the sonar system provides the precise location of the vessel for each sonar measurement. The GNSS+INS system also provides the vertical location of the vessel to compensate for waves.

Unmanned Vehicles

An unmanned vehicle is a vehicle that is unoccupied but under human control, whether radio-controlled or automatically guided by a GNSS-based application. There are many types of unmanned vehicles, including: Unmanned Ground Vehicle (UGV), Unmanned Aerial Vehicle (UAV), Unmanned Surface Vehicle (USV) and Unmanned Underwater Vehicle (UUV).

Predator unmanned aerial vehicle (UAV).

Initially, unmanned vehicles were used primarily by the defense industry. However, as the unmanned vehicle market has grown and diversified, the commercial use of unmanned vehicles has also grown and diversified. Some of the current civilian uses for unmanned vehicles are: search and rescue, crop monitoring, wildlife conservation, aerial photography, environmental research, infrastructure inspection, bathymetry, landmine detection and disposal, HAZMAT inspection and disaster management. As the civilian unmanned vehicle market expands, so will the civilian use of unmanned vehicles.

Hurricane Research

Knowing where a hurricane will make land fall and how powerful it will be are important to properly prepare for the storm. While meteorologists are good at predicting the potential path of a hurricane it is much harder to predict how powerful the storm will be when it arrives.

To learn more about what causes a hurricane to rapidly increase or decrease in intensity, NASA is using two long range UAVs to study the storms while they are still far out to sea. Onboard the UAVs are meteorological instruments that monitor the environmental conditions in and around the storm. The UAVs also have a GNSS+INS system that records the UAV location and attitude for each measurement taken by the meteorological instruments. An accurate UAV location and attitude is necessary for the measurement take to be useful.

Orion Spacecraft Parachute Testing

Before the Orion spacecraft can be used for manned space missions, NASA must know that Orion can safely land on earth. A key aspect of returning the astronauts safely to earth is slowing the Orion spacecraft from its incredibly high reentry speed of close to 32,000 km/h to less than 36 km/h. This is the job of the Capsule Parachute Assembly System (CPAS).

To test the CPAS, NASA created two unmanned test vehicles. These test vehicles are dropped out of a C-17 aircraft from altitudes as high as 35,000 feet. A GNSS+INS system installed in the test vehicles measures the vertical velocity of vehicle to test the effectiveness of parachute system.

Landing an Unmanned Helicopter on a Ship

The autonomous landing of an unmanned helicopter is already challenging as the navigation system has to deal with the movement of the helicopter caused by the winds. This challenge is greatly increased when trying to land on a ship at sea. Not only is the helicopter's position changing based on its movement and the effects of the wind, the ship is moving independently based on its movement and the effects of both the wind and the sea.

When landing a helicopter on a ship, the relative distance between the helicopter landing gear and the flight deck of the ship is much more important than the absolute position of the helicopter and ship. GNSS+INS systems installed on both the helicopter and the ship are used to determine this relative distance. The GNSS+INS system on the ship calculates its position and sends that information to the GNSS+INS system on the helicopter. The GNSS+INS system on the helicopter uses the position sent from the ship along with its own position to calculate the relative distance and direction between the ship and helicopter. Using this relative distance and direction, the unmanned helicopter is able to autonomously approach and land on the ship's flight deck.

Defence

The defence sector makes broad use of GNSS technology, including:

- Navigation: Using GNSS receivers, soldiers and pilots can navigate unfamiliar terrain or conduct night-time operations. Most foot soldiers now carry handheld GNSS receivers.

- Search and Rescue: If a plane crashes and that plane has a search and rescue beacon that is equipped with a GNSS receiver, it may be possible to more quickly locate it.
- Reconnaissance and Map Creation: The military uses GNSS to create maps of uncharted or enemy territory. They can also mark reconnaissance points using GNSS.
- Unmanned Vehicles: Unmanned vehicles are used extensively in military applications, including reconnaissance, logistics, target and decoy, mine detection, search and rescue, research and development and missions in unsecured or contaminated areas.
- Munitions Guidance: Precision munitions use GNSS to ensure the munition lands on target.

Automotive Navigation Systems

Automotive navigation systems are a satellite navigation system which is installed in automobiles. Automotive navigation systems use GPS navigation devices to obtain position data to locate the users on road in the units map database. With the help of road data base, the unit gives instructions to other locations along the road. The enhancement in automotive navigation systems technology has given growth to a number of gadgets and devices. APAC Automotive navigation systems are witnessing a rapid growth in adoption of such devices. For instance, the Indian automobile industry has been witnessing growth over few decades. Owing to the growing competition in the Indian Automotive navigation systems market, competitors are equipped with latest technological invention. In India the first automobile navigation system was launched by SatNav Technologies followed by MapyIndia and Blaupunkt. One of the recent innovations was by Blaupunkt. Their GPS device is called TravelPoint. It is a portable device available in various models. All the models provide direction based on a combination of verbal and visual communication.

Penetration of automobile MEMS sensor is one of the emerging applications and is being used in electronic control unit, and tire pressure monitoring systems in automotive navigation systems market. Owing to the increasing security concerns in vehicle across the globe, the demand for automotive MEMS sensors are surging. The major market for MEMS sensors are in European and American countries.

Automotive Navigation Systems Market Segmentation

Automotive navigation systems market can be segmented by vehicle types into:

- Entry-level passenger vehicle.

- Mid-premium passenger vehicle.
- Luxury passenger vehicle.
- Commercial vehicle.

On the basis of navigation system automotive navigation systems market can be further segmented into:

- Aftermarket IVS.
- Factory fitted IVS.
- Personal navigation device (PND's).
- Smartphones/Tablets.

Almost 81% of the PND's sold in 2015 automotive navigation systems market had integrated cellular connectivity. Europe Automotive navigation systems market is one of the major markets in PND's.

However, North America Automotive navigation systems market is gaining traction and the shipments have increased in the region. Automotive navigation systems markets such as Europe and North America where the penetration rates are already high, PND as a device category is facing increasing competition from handset-based navigation services and availability of low cost in-dash car navigation systems. However new services and features in automotive navigation systems are becoming more important as an increasing share of PND sales come from replacement devices for existing PND users.

The integration of smartphones with IVS (in-vehicle systems) is one of the primary reasons for the growth in automotive navigation systems market. In the recent years, adoption of smartphone has increased and this is expected to surge the demand for smartphone/tablet based navigation devices. Increased consumer demand for wireless connectivity and mobile internet with high level of penetration of smartphones and tablets will place connected cars first in demand. This is expected to drive the supply side vendors in automotive industry to adjust to new digital lifestyle and to integrate in-care internet connectivity solutions hence rising demand of automotive navigation systems.

Technological advancements are also driving market for automotive navigation systems market. For instances, the devices with integration of traffic data with navigation systems is one such innovation. These newer systems give precise driving directions and receive and display information on traffic congestion and suggest alternative routes to the users. Either TMC, which delivers coded traffic information using RDS is used or GPRS/3Gdata transmitted via mobile phones.

The other functions and integrations in the devices are expected to enhance the sales for automotive navigation systems. For instance, the color LCD screens on few automobiles navigation systems can also be used as television broadcasts or DVD movies. Few of the systems can also be integrated with mobile phones for hands free talking and SMS messaging using Bluetooth and Wifi. Nowadays Automotive navigations systems include personal management for meetings and can be combined with traffic and public transport information system.

References

- Satellite-navigation, definition-30440: techopedia.com, Retrieved 02 March, 2020
- What-is-gnss: oxts.com, Retrieved 18 June, 2020
- GNSS-Reflectometry: mistrale.eu, Retrieved 25 July, 2020
- Applications,an-introduction-to-gnss-chapter-8-gnss-applications-and-equipment:novatel.com, Retrieved 16 August, 2020
- Automotive-navigation-systems-market, market-research: persistencemarketresearch.com, Retrieved 28 January, 2020

2

Global Navigation Satellite System Augmentation

The method that is used for the improvement of navigation system's attributes like accuracy, reliability and availability, through the use of external resources is called GNSS augmentation. It has two major systems namely local area and wide area augmentation system. This chapter closely examines the method of GNSS augmentation to provide an extensive understanding of the subject.

Augmentation of a global navigation satellite system (GNSS) is a method of improving "augmenting" the navigation system's performances, such as integrity, continuity, accuracy or availability thanks to the use of external information to the GNSS into the user position solution.

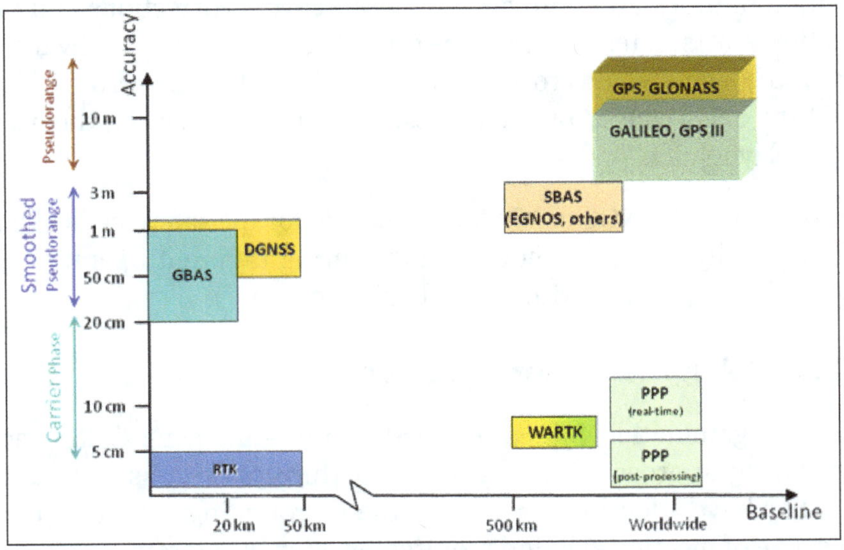

Accuracy performances for GNSS and GNSS augmentation techniques.

Based on its main feature, GNSS augmentation systems can be classified as those providing integrity information to the primary GNSS satellites constellations and those improving the accuracy of the user solution with respect to the only use of the primary

GNSS constellations. A further classification may be done according to an additional relevant feature, which for the former relates on whether the augmentation information comes from satellites (satellite-based augmentation system) or from ground (ground-based augmentation system) and for the latter on whether the accuracy improvements use a dense network of reference stations (Differential GNSS, Real Time Kinematics RTK or Wide Area RTK (WARTK)) or just a few stations (Precise Point Positioning PPP) for the computation of the augmentation information.

Satellite-based Augmentation Systems

A Satellite-based Augmentation System (SBAS) is a civil aviation safety-critical system that supports wide-area or regional augmentation even continental scale through the use of geostationary (GEO) satellites which broadcast the augmentation information. A SBAS augments primary GNSS constellations by providing GEO ranging, integrity and correction information. While the main goal of SBAS is to provide integrity assurance, it also increases the accuracy with position errors below 1 metre (1 sigma).

The ground infrastructure includes the accurately-surveyed sensor stations which receive the data from the primary GNSS satellites and a Central Processing Facility (CPF) which computes integrity, corrections and GEO ranging data forming the SBAS signal-in-space (SIS). The SBAS GEO satellites relay the SIS to the SBAS users which determine their position and time information. For this, they use measurements and satellite positions both from the primary GNSS constellations and the SBAS GEO satellites and apply the SBAS correction data and its integrity.

The augmentation information provided by SBAS covers corrections and integrity for satellite position errors, satellite clock time errors and errors induced by the estimation of the delay of the signal while crossing the ionosphere. For the errors induced by the estimation of the delay caused by the troposphere and its integrity, the user applies a tropospheric delay model.

SBAS systems, such as EGNOS or WAAS, are usable for the safety-critical task of guiding aircraft -vertically as well as horizontally- during different operations, including landing approaches (approach with vertical guidance, APV).

Ground-based Augmentation Systems

Ground-based Augmentation System (GBAS) aim at enhancing GNSS service levels for aviation during approach, landing and departure phases, as well as for surface operations. They have a local coverage (e.g. the surroundings of an airport) with the primary objective of meeting aviation requirements for the aforementioned operations and phases, in terms of accuracy, integrity and safety. Actually, whereas the main goal of GBAS is to provide integrity assurance, it also increases the accuracy with position errors below 1 m (1 sigma).

The ground infrastructure includes two or more GNSS receivers which collect pseudoranges for all the primary GNSS satellites in view and computes and broadcasts differential corrections and integrity-related information for them based on its own surveyed position. These differential corrections are transmitted from the ground system via a Very High Frequency (VHF) Data Broadcast (VDB). The broadcast information includes pseudorange corrections, integrity parameters and various locally relevant data such as Final Approach Segment (FAS) data, referenced to the World Geodetic System (WGS-84). Any aircraft (within the area of coverage of the ground station) may use those corrections to compute their own measurements to compute a (differentially corrected) position, which at the same time is used to generate navigation guidance signals.

GBAS and SBAS are both GNSS safety critical systems for civil aviation which share similar principles. The main difference comes from the fact that GBAS provides local corrections to the satellite pseudoranges using just ground infrastructure in the vicinity of the served airport, whilst SBAS broadcasts corrections to the different components of the pseudorange error valid for an area as big as a continent; the price to pay is that the SBAS infrastructure needs tens of sensor distributed in the augmented area and two or more GEO satellites to broadcast the information.

Differential GNSS

Differential GNSS is an enhancement to primary GNSS constellations information by the use of a network of ground-based reference stations which enable the broadcasting of differential information to the user also named rover to improve the accuracy of his position the integrity is not assured. There are several differential GNSS techniques, such as the classical DGNSS (or DGPS), the Real Time Kinematics (RTK) and the Wide Area RTK (WARTK).

Classical DGNSS

The DGNSS technique consists of the determination of the GNSS position for an accurately-surveyed position known as reference station. Given that the position of the reference station is accurately known, the deviation of the measured position to the actual position and more importantly the corrections to the measured pseudoranges to each of the individual satellites can be calculated. These corrections can thereby be used for the correction of the measured positions of other GNSS user receivers.

DGNSS accuracy is in the order of 1 m (1 sigma) for users in the range of few tens of km from the reference station, growing at the rate of 1 m per 150 km of separation.

Real Time Kinematic

RTK is a technique widely used for precise GNSS positioning based on the use of code and carrier phase measurements from the primary GNSS constellations. The use of carrier phase measurements allows reaching cm-level accuracies at the expense of having to solve the ambiguity of such carrier phases, which is a sophisticated process with a certain convergence time.

The main inconvenience of the RTK technique is that it requires a reference station relatively close to the user so that the differential ionospheric delay is negligible. A way to partially overcome such inconvenience appears with the network RTK (NRTK or virtual RTK) which uses a set of reference stations to mitigate atmospheric dependent effects, such as the ionospheric delay, over distance. In any case, RTK/NRTK approaches works well with baselines no longer than about 15 km.

Wide Area RTK

The Wide Area RTK (WARTK) technique was introduced in the late 1990s to overcome the RTK need of a very dense reference stations network. The WARTK concept provides accurate ionospheric corrections that are used as additional information and allows increasing the RTK service area. In this way, it just needs reference stations separated by about 500–900 kilometres. WARTK is a relatively new concept with still no operational system in place.

Precise Point Positioning

Precise Point Positioning (PPP) is another method for precise position determination. PPP requires the availability of precise reference GNSS orbits and clocks in real-time. Combining the precise satellite positions and clocks with a dual-frequency GNSS receiver (to remove the first order effect of the ionosphere), PPP is able to provide position solutions at centimetre to decimetre level.

PPP differs from double-difference RTK positioning in the sense that it does not require access to observations from one or more close reference stations accurately-surveyed. PPP just requires data from reference stations from a relatively sparse station network (thousands of km apart would suffice). This makes PPP a very attractive alternative to RTK for those areas where RTK coverage is not available. On the contrary, the PPP technique is still not so much consolidated as RTK and requires a longer convergence time to achieve maximum performances (in the order of tenths of minutes).

Satellite based Augmentation Systems

A Satellite-based Augmentation System (SBAS) is a civil aviation safety-critical system that supports wide-area or regional augmentation even continental scale through the use of geostationary (GEO) satellites which broadcast the augmentation information. A SBAS augments primary GNSS constellations by providing GEO ranging, integrity and correction information. While the main goal of SBAS is to provide integrity assurance, it also increases the accuracy with position errors below 1 meter (1 sigma).

The ground infrastructure includes the accurately-surveyed sensor stations which receive the data from the primary GNSS satellites and a Processing Facility Center which computes integrity, corrections and GEO ranging data forming the SBAS signal-in-space (SIS). The SBAS GEO satellites relay the SIS to the SBAS users which determine their position and time information. For this, they use measurements and satellite positions both from the primary GNSS constellations and the SBAS GEO satellites and apply the SBAS correction data and its integrity.

The augmentation information provided by SBAS covers corrections and integrity for satellite position errors, satellite clock/time errors and errors induced by the estimation of the delay of the signal while crossing the ionosphere. For the errors induced by the estimation of the delay caused by the troposphere and its integrity, the user applies a tropospheric delay model.

SBAS Signal

Every SBAS provides ranging signals transmitted by GEO satellites, differential corrections on the wide area and additional parameters aimed to guarantee the integrity of the GNSS user:

- GEO Ranging: Transmission of GPS-like L1 signals from GEO satellites to augment the number of navigation satellites available to the users.
- Wide Area Differential (WAD): Differential corrections to the existing GPS/GLONASS/GEO navigation services computed in a wide area to improve navigation services performance. This includes corrections to the satellite orbits and clocks, as well as information to estimate the delay suffered from the signal when it passes through the ionosphere.
- GNSS/Ground Integrity Channel (GIC): Integrity information to inform about the availability of GPS/GLONASS/GEO safe navigation service.

The SBAS delivers to the user the corrections and integrity data as well as some ancillary information (timing, degradation parameters, etc.) through messages encoded in the signal. The specification of the SBAS message data format is contained in the ICAO SARPS Appendix B for the aspects related with the signal in space, as well as in the RTCA MOPS DO-229D for the minimum performance requirements applicable to the airborne SBAS receiver equipment.

The SBAS satellite shall transmit a GPS-like L1 (1574.42 MHz) signal, modulated with a Coarse/Acquisition Pseudo-Random Noise (PRN) code. The SBAS L1 radiofrequency characteristics are.

Parameter	Description
Modulation	Bi-phase shift key (BPSK) modulated by a bit train comprising the PRN code and the SBAS data (modulo-2 sum).
Bandwidth	L1 ±30.69 MHz. At least 95% of the broadcast power will be contained within the L1 ±12 MHz band.
Ranging Codes	A PRN Code (Gold code) of 1 millisecond in length at a chipping rate of 1023 Kbps.
SBAS Data	500 symbols per second.
Power	Minimum power −131 dBm at 5 degrees elevation Maximum power −119,5 dBm.

A convolutional encoding of the bits is performed with the following parameters:

Code Parameter	Value
Coding Rate	1/2
Coding Scheme	Convolutional
Constraint Length	7
Generator Polynomials	G1 = 171(oct); G2 = 133(oct)
Encoding Sequence	G1 then G2
Flush	No

SBAS Performances

The SBAS performances are defined with respect to the level of service that the system is designed to provide. The main source for SBAS performances comes from civil aviation navigation safety requirements and they are different for each civil aviation operation.

Typical Operation	Horizontal Accuracy (95%)	Vertical Accuracy (95%)	Integrity	Time-To-Alert (TTA)	Continuity	Availability
En-route	3.7 km (2.0 NM)	N/A	1 − 1 × 10-7/h	5 min	1 − 1 × 10-4/h to 1 − 1 × 10-8/h	0.99 to 0.99999
En-route Terminal	0.74 km (0.4 NM)	N/A	1 − 1 × 10-7/h	15 s	1 − 1 × 10-4/h to 1 − 1 × 10-8/h	0.99 to 0.99999
Initial approach, Intermediate approach, Non-precision approach (NPA), Departure	220 m (720 ft)	N/A	1 −1x10-7/h	10 s	1 − 1x10-4/h to 1 − 1x10-8/h	0.99 to 0.99999

Approach operations with vertical guidance (APV-I)	16 m (52 ft)	20 m (66 ft)	1 – 2 × 10-7 per approach	10 s	1 – 8 × 10-6 in any 15 s	0.99 to 0.99999
Approach operations with vertical guidance (APV-II)	16 m (52 ft)	8 m (26 ft)	1 – 2 × 10-7 per approach	6 s	1 – 8 × 10-6 in any 15 s	0.99 to 0.99999
Category I precision Approach	16 m (52 ft)	6.0 m to 4.0 m (20 ft to 13 ft)	1 – 2 × 10-7 per approach	6 s	1 – 8 × 10-6 in any 15 s	0.99 to 0.99999

As indicated in the table above, the performance requirements are expressed in terms of four quantitative concepts, many of them to be interpreted as probabilistic figures:

- Accuracy: Is expressed in terms of Navigation System Error (NSE) as the difference between the real position of the aircraft and the position provided by the airborne equipment. A SBAS assures the compliance with respect the accuracy requirements by providing to the user corrections to the satellite orbit and clock errors as well as to the ionospheric residual propagation error. For LPV200 operations, new requirements have been proposed distinguishing the fault-free and the system failure conditions.

- Integrity: Is defined by ICAO as a measure of the trust that can be placed in the correctness of the information supplied by the system. This general statement is expressed at the SBAS system level as the maximum allowable probability that the navigation position error exceeds alarm limit and the navigation system does not alert the pilot in a time less than the time to alert. The SBAS assures the integrity requirements by:

    - Providing to the user satellite and/or ionospheric alarms in order to inform the user to reject the corresponding satellite/ionospheric corrections in its positioning computation.

    - Providing to the user Horizontal and Vertical Protection Level information (HPL, VPL) in order to assess the availability of the system, by comparing these PLs with the corresponding Alarm Limits (AL) for a given phase of flight. The SBAS computes and broadcasts integrity bounds to the satellite orbit and clock (UDRE) corrections as well as to the ionospheric corrections errors (GIVE) so that the user is able to compute a PL that over bounds the navigation system error experienced by the user with the integrity risk requirement.

- Continuity: Is the probability that the specified system performance will be maintained for the duration of a phase of operation, presuming that the system was available at the beginning of that phase of operation and was predicted to operate throughout the operation. Lack of continuity means that the operation must be aborted (with the associated risk).

- Availability: Is the probability that the navigation service is available at the beginning of the planned operation. A SBAS is considered available when the accuracy, integrity and continuity requirements are met and it is measured in terms of probability of the system being available for any given user at any given time. In practice, the availability is computed by measuring the probability of a protection level being below its corresponding alarm limit. It should be noted that a lack of availability is not a safety concern but prevents the nominal operation of the system, and implies an associated impact on the service operation status.

Operation	Horizontal AL	Vertical AL
En-route (oceanic/continental)	7.4 Km (4 NM)	N/A
En-route (continental)	3.7 Km (2 NM)	N/A
En-route, Terminal	1.85 Km (1 NM)	N/A
NPA	556 m (0.3 NM)	N/A
APV-I	40 m (130 ft)	50 m (164 ft)
LPV200	40 m (130 ft)	35 m (200 ft)
APV-II	40 m (130 ft)	20 m (66 ft)
Category I	40 m (130 ft)	15 to 10 m (50 ft to 33 ft)

SBAS Architecture

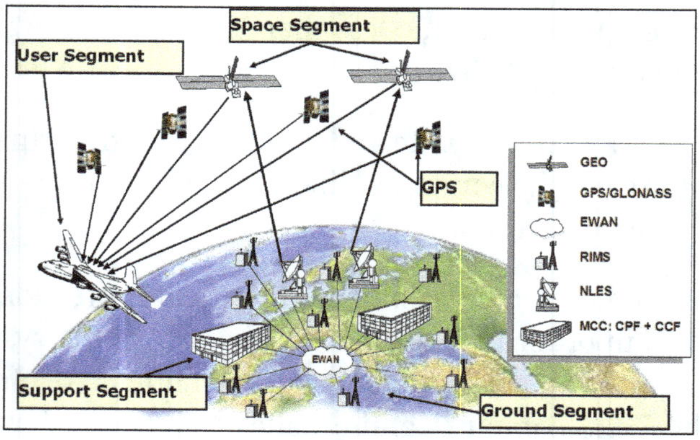

SBAS architecture.

A SBAS is a safety critical system designed to augment one or several satellite navigation systems. In this sense, it is not a standalone system. The main layers of a general SBAS architecture are:

- Space segment comprising the geostationary satellites (GEO) with navigation payloads in charge of transmitting a GPS-like carrier signal with the SBAS information.

- Ground segment comprising all the ground elements in charge of the provision of the SBAS navigation message. The main elements are:
    - Monitoring Station Network.
    - Processing Facility Center.
    - GEO Satellite Control Center.
    - Communication Layer.
- Support segment comprising all the elements need to support the correct operation and maintenance of the SBAS: Configuration control, performance evaluation, maintenance and development, help desk, etc.
- User segment comprising all the user equipment needed to receive and use the SBAS information.

Space Segment

The SBAS space segment is composed by several geostationary satellites in charge of broadcasting, over the service area, the SBAS navigation message. Typically, the SBAS satellites are multi-purpose (commercial communication) satellites that carry out an additional navigation payload capable to generate a GPS-like signal that retransmits to the users the navigation message generated on-ground.

The SBAS GEO navigation payload is a transponder that relays the on-ground generated signal (upload to the GEO in the C-band frequency) to a transmitted signal in the L-band frequency band. For most of the navigation payloads, a second C-band downlink channel is available in order to improve the adjustment of the signal delay due to atmospheric propagation. However, first generation of navigation payloads only included a single-channel frequency translating repeater.

The new generation of SBAS navigation payload includes more advanced functionalities. Among them, it is worth noting:

- Increase of the emission bandwidth and broadcast power.
- Evolution to dual-frequency L1/L5 transponders.
- Evolution of payload from signal transponders to regenerative payloads in which the signal is entirely build up onboard (the GEO only receives from ground segment the navigation message to be modulated in the GEO signal).

Ground Mission Segment

The main purpose of the Ground Mission Segment is to generate and uplink the augmentation signal that will be broadcast by the GEO satellite.

To achieve this objective, the Ground Mission Segment is typically broken down into the following subsystems:

- Monitoring Station Network: Its purpose is to collect data from the satellites that are to be augmented. This is performed by a network of GNSS receivers.
- Processing Facility Center: It is in charge of processing the data provided by the Monitoring Station Network to generate the messages to be broadcasted to the satellites.
- GEO Satellite Control Center: It is in charge of generating the signal with the message provided by the Processing Facility Center and up-linking it to the GEO satellites.
- Communication Layer: It interconnects the different elements of the Ground Segment.

Monitoring Station Network

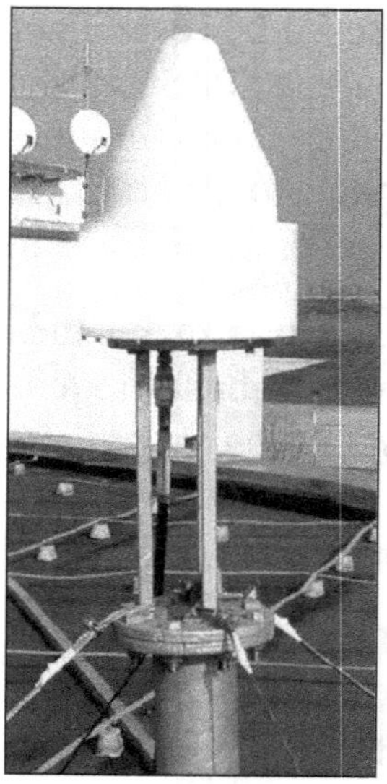

EGNOS reference station.

The Monitoring Station Network is carefully designed to monitor satellites and ionosphere with the required accuracy and availability in the SBAS service area. This usually implies a dense network of stations in the service area, with some additional stations outside. In addition, there must be redundancy to avoid single points of failure, or even different subsets of stations to feed parallel and independent processing chains.

The monitoring stations have the following characteristics:

- Dual frequency (L1/L2) receivers with geodetic quality.
- Atomic frequency standard (Caesium/Rubidium/H-Maser).

- Able to track all in view GPS and GEO satellites.
- Allocated in site conditions with good local environment condition in terms of multipath and radio frequency interference.
- Geo-referenced to WGS-84 (or ITRF) within a 1-3 cm accuracy.
- Compliant with some processing capabilities:
  - 1 Hz data acquisition.
  - Embedded data quality checks to remove misleading data.
  - Data processing and broadcast within a few milliseconds.
  - Integrated robustness against known threats (e.g. *evil* waveforms detection).

Processing Facility Center

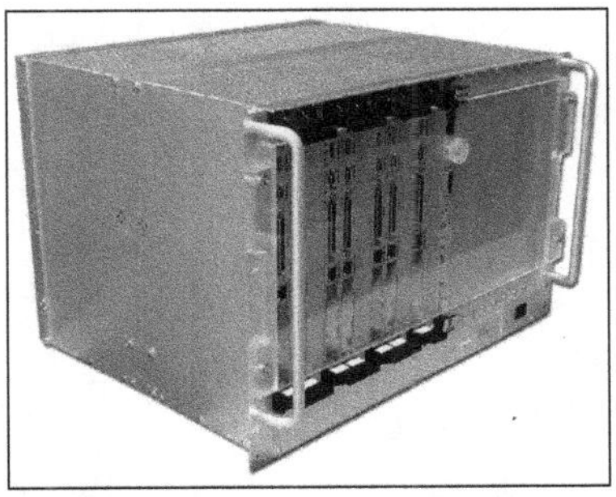

CPFPS, component of the EGNOS Central Processing Facility.

The Processing Facility Center is one of the critical elements of the SBAS system as it is charge of generating the augmentation information for the user, playing a major role in the achievement of the SBAS performance requirements.

The Processing Facility Center is in charge of the following functions:

- Process all input data from the Monitoring Station Network.
- Estimate the satellite corrections, ionospheric model and error variance terms.
- Perform a dedicated integrity assessment on the information provided by the system to the users.
- Format the outputs according to the SBAS standards.

The Processing Facility Center has a very strong algorithmic load, as it has to solve the following challenges:

- Advanced models for satellite orbit determination (breaking frontiers in GEO orbit and clock determination).

- Determination of the SBAS reference time and satellite clock corrections with better than 2 ns accuracy.
- Real-time ionospheric estimation with a dense network of receivers. This estimation should be able to characterize local and rapid phenomena.
- Integrity estimation valid for wide-areas with demanding performance requirements.

GEO Satellite Control Center

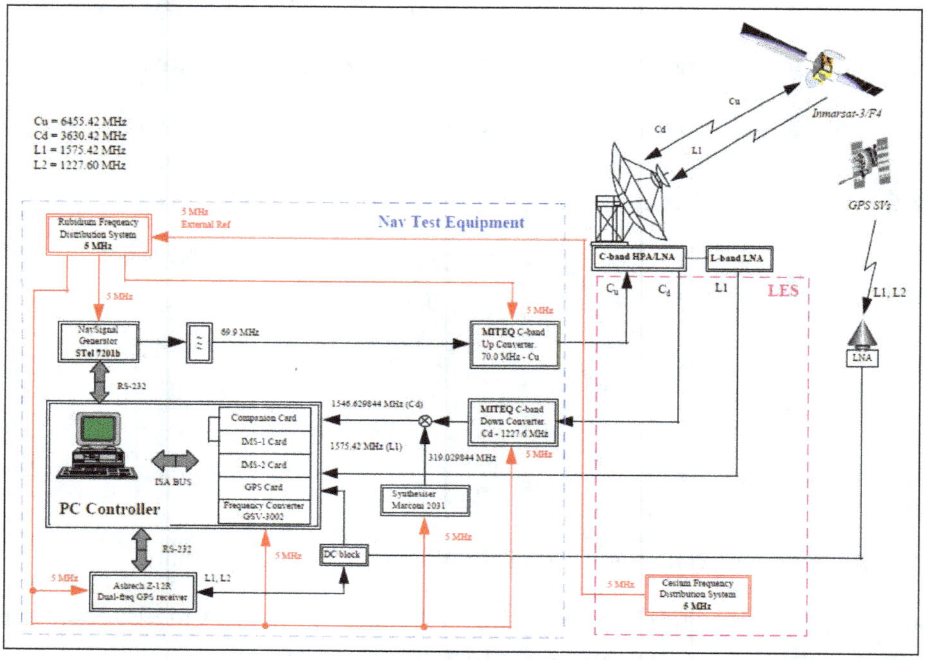

Test diagram for EGNOS uplink station.

GEO Satellite Control Center function is to relay the SBAS messages computed by the Processing Facility Center to the GEO satellites. Its functions include:

- Codify the GPS-like signal encoding. In particular, the preparation of the signal generation as well as the PRN encoding.
- Modulate the SBAS message information provided by the Processing Facility Center into the signal.
- Synchronize the timing of the signal to the SBAS reference time.
- Control de time synchronization between the carrier and code components of the uplink signal (core-carrier coherency).
- Close the uplink-station/GEO-satellite control loop by receiving and processing the downlink SBAS satellite signals (both in L- and C- bands).

Communication Layer

The Communication Layer ensures that all the SBAS Ground Segment elements are interconnected in real time. This connection should be performed with high performance standards in terms of reliable lines, sufficient bandwidth for a large exchange of data, and stringent redundancy and security requirements.

Support Mission Segment

The SBAS Support Segment collects all the elements need to support the development and operation of a SBAS system. These elements are not related with the provision of the SBAS service but they are needed as external support facilities. They SBAS support needs can be listed as:

- The design, development and validation phases.
- The deployment and operation.
- The certification process.
- The maintenance and troubleshooting.
- The certification of applications.

Between the different SBAS Support elements it is worth noting:

- End to End Simulators in charge of simulating data under controlled conditions (reference errors, emulated feared events) in order to simulate the real system behaviour (delays and communication problems, algorithm, broadcasting features, etc).
- Service Volume Simulators in charge of evaluating the overall SBAS performances over the service area under a set of controlled conditions.
- Performance Analysis Tools in charge of measuring all the performance concepts at:
    - System level by analyzing the Signal in Space information and checking the accuracy, integrity, continuity and availability features but using concepts based in the errors available at pseudorange domain.
    - User point of view level, by analyzing user data (including local conditions) and applying user-related algorithms.
- On-line monitoring tools for tracking the system behavior in real time and collecting system failure and anomalies. These elements also support the system operation by predicting the expected behavior of the system and provide warnings as system outages or degradations. They also provide warnings to external users.

- Archiving tools supporting the data archiving. The data comprises both all the internal information generated by the system as well as external data (satellite reference orbit and clock data, ionospheric information, outages, etc).
- Application test benches to support applications by generating test scenarios for their certification and performance verification. They are able to plug different application software or equipment for homologation purposes.

User Segment

The SBAS user segment comprises all the user equipment that makes use of the SBAS Signal in Space (SIS). In fact, the SBAS User segment is not under the control of the SBAS service provider as it is driven by the SBAS application market.

In general, the SBAS service operator provides different services aiming at different market sectors, namely an Open Service, a Safety of Life service (SoL) and even a Commercial Service.

For the Safety-of-Life (SoL) service, the SBAS user equipment shall be compliant (certified) against several standards. For instance, civil aviation SBAS equipment shall demonstrate:

- Full compliance to RTCA SBAS MOPS DO-229 (airborne equipment).
- Full compliance to the RTCA SBAS MOPS 228 and 301 (antenna requirements).
- Compliance to RTCA TSO (C190, C145b, C146b) for SBAS equipment.
- Compatibility to other avionics equipment, in particular Flight Management Systems (FMS).

The SoL civil aviation certified equipment is in the highest rank with respect its cost. There exist a large number of certified receivers manufacturers worldwide both in the US (GARMIN, Honeywell, Rockwell Collins, General Avionics, etc) and in Europe.

The Open Service (OS) targets low cost, general purpose GNSS equipment that uses the SBAS SIS to provide the user with an enhanced accuracy performance in comparison with the one provided by a standalone GPS device. In comparison with the certification requirements of the user equipment above, user equipment is not necessarily compliant with the RTCA MOPS DO 229 processing rules, but might only make use of the processing algorithms that render the accuracy corrections provided by the SBAS SIS.

Finally, some SBAS service providers include the provision of the information computed by the SBAS ground element (input raw data, corrections, integrity information) by dissemination means different to the SBAS GEO link (generally, using terrestrial telecommunication networks). This constitutes the Commercial Service. This market sector comprises professional users (land and geodesy applications, maritime or terrestrial transport applications) that are not subject to the integrity and latency requirements needed in the SoL service.

European Geostationary Navigation Overlay Service

The European Geostationary Navigation Overlay Service (EGNOS) is a satellite based augmentation system (SBAS) developed by the European Space Agency and EUROCONTROL on behalf of the European Commission. Currently, it supplements the GPS by reporting on the reliability and accuracy of their positioning data and sending out corrections. The system will supplement Galileo in a future version.

EGNOS consists of a network of about 40 ground stations and 3 geostationary satellites. Ground stations determine accuracy data of the satellite navigation systems and transfer it to the geostationary satellites; users may freely obtain this data from those satellites using an EGNOS-enabled receiver, or over the internet. One main use of the system is in aviation.

According to specifications, horizontal position accuracy when using EGNOS-provided corrections should be better than seven metres. In practice, the horizontal position accuracy is at the metre level.

Similar service is provided in North America by the Wide Area Augmentation System (WAAS), and in Asia, by Japan's Multi-functional Satellite Augmentation System (MSAS) and India's GPS Aided GEO Augmented Navigation (GAGAN).

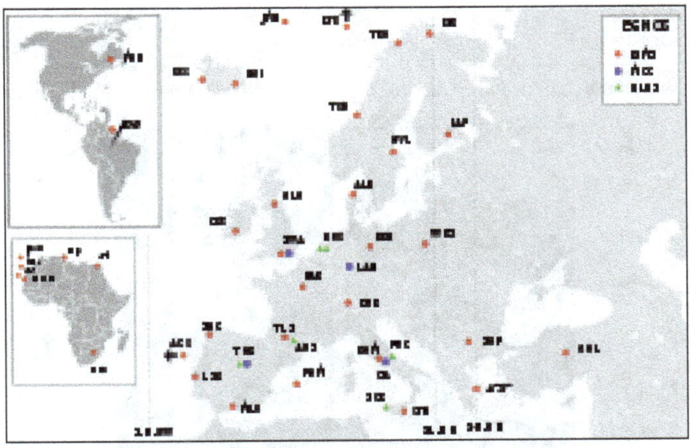

Map of the EGNOS ground network.

Satellites and SISNeT

Satellite Name and Details	NMEA/PRN	Signals	Location	Status
Inmarsat 3-F2 (Atlantic Ocean Region-East)	NMEA #33/PRN #120	L1	15.5°W	retired
ARTEMIS	NMEA #37/PRN #124	-	21.5°E	retired

Inmarsat 4-F2 (Europe Middle East Africa)	NMEA #39/PRN #126	-	64°E	testing
Inmarsat 3-F1 (Indian Ocean)	NMEA #44/PRN #131	-	64.5°E	retired
SES-5 (a.k.a. Sirius 5 or Astra 4B)	NMEA #49/PRN #136	L1 & L5	5.0°E	active
Astra 5B	NMEA #36/PRN #123	L1 & L5	31.5°E	active

Similar to WAAS, EGNOS is mostly designed for aviation users who enjoy unperturbed reception of direct signals from geostationary satellites up to very high latitudes. The use of EGNOS on the ground, especially in urban areas, is limited due to relatively low elevation of geostationary satellites: about 30° above horizon in central Europe and much less in the North of Europe. To address this problem, ESA released in 2002 SISNeT, an Internet service designed for continuous delivery of EGNOS signals to ground users. The first experimental SISNeT receiver was created by the Finnish Geodetic Institute. The commercial SISNeT receivers have been developed by Septentrio. PRN #136 was placed into the Operational Platform from 23/08/2018 at 10:00 UTC and PRN #120 was placed into Test Platform from 30/08/2018 at 13:00 UTC.

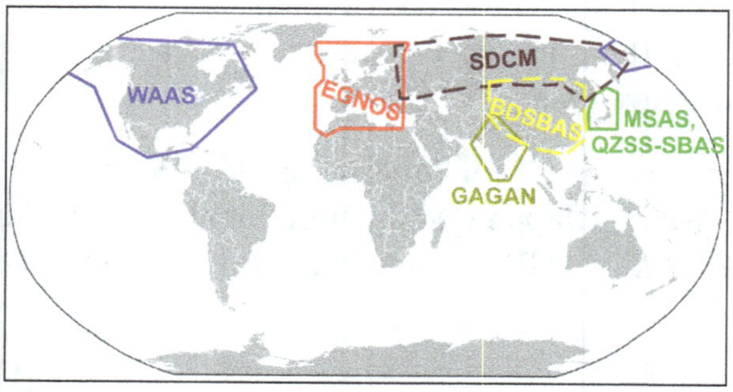

Service areas of satellite-based augmentation systems (SBAS).

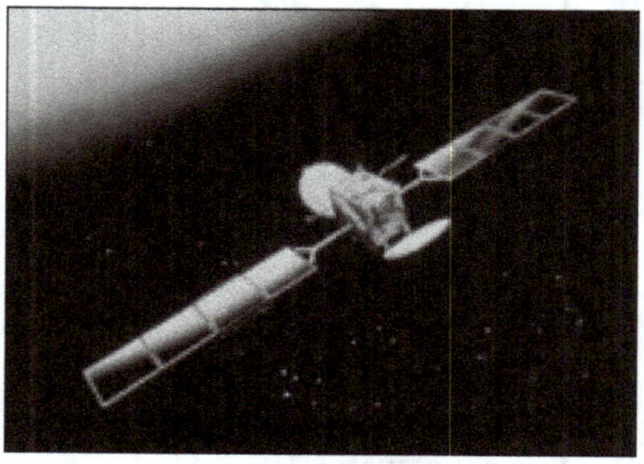

Inmarsat-3-Satellite.

Ground Stations

More than 40 ground stations are linked together to create EGNOS network which consists:

- 40 ranging and integrity monitoring stations (RIMS): Receiving signals from US GPS satellites.

- 4 mission control centers (MCC): Data processing and differential corrections counting.
- 6 navigation land earth stations (NLES): Accuracy and reliability data sending to three geostationary satellite transponders to allow end-user devices to receive them.

EGNOS RIMS "BRN".

Aviation

In March 2011, the EGNOS Safety-of-Life Service was deemed acceptable for use in aviation. This allows pilots throughout Europe to use the EGNOS system as a form of positioning during an approach, and allows pilots to land the aircraft in IMC using a GPS approach.

As of September 2014 LPV (Localizer performance with vertical guidance) landing procedures, which are EGNOS-enabled, were available at more than 114 airports across Europe.

GPS-aided GEO Augmented Navigation

The GPS-aided GEO augmented navigation (GAGAN) is an implementation of a regional satellite-based augmentation system (SBAS) by the Government of India. It is a system to improve the accuracy of a GNSS receiver by providing reference signals. The AAI's efforts towards implementation of operational SBAS can be viewed as the first step towards introduction of modern Communication, navigation and surveillance/Air Traffic Management system over Indian airspace.

The project has established 15 Indian reference stations, 3 Indian navigation land uplink stations, 3 Indian mission control centers, and installation of all associated software and communication links. It will be able to help pilots to navigate in the Indian airspace by an accuracy of 3 m. This will be helpful for landing aircraft in marginal weather and difficult approaches like Mangalore and Leh airports.

Implementation

The 7.74 billion (US$109 million) project was created in three phases through 2008 by the Airport Authority of India with the help of the Indian Space Research Organisation's (ISRO) technology and space support. The goal is to provide navigation system for all phases of flight over the Indian airspace and in the adjoining area. It is applicable to safety-to-life operations, and meets the performance requirements of international civil aviation regulatory bodies.

The space component became available after the launch of the GAGAN payload on the GSAT-8 communication satellite, which was successfully launched. This payload was also part of the GSAT-4 satellite that was lost when the geosynchronous satellite launch vehicle (GSLV) failed during launch in April 2010. A final system acceptance test was conducted during June 2012 followed by system certification during July 2013.

Technology

To begin implementing a satellite-based augmentation system over the Indian airspace, Wide Area Augmentation System (WAAS) codes for L1 frequency and L5 frequency were obtained from the United States Air Force and U.S Department of Defense on November 2001 and March 2005. The system will use eight reference stations located in Delhi, Guwahati, Kolkata, Ahmedabad, Thiruvananthapuram, Bangalore, Jammu and Port Blair, and a master control center at Bangalore. US defense contractor Raytheon has stated they will bid to build the system.

Technology Demonstration

A national plan for satellite navigation including implementation of technology demonstration system (TDS) over the Indian air space as a proof of concept had been prepared jointly by Airports Authority of India (AAI) and ISRO. TDS was successfully completed during 2007 by installing eight Indian Reference Stations (INRESs) at eight Indian airports and linked to the Master Control Center (MCC) located near Bangalore. Preliminary system acceptance testing has been successfully completed in December 2010. The ground segment for GAGAN, which has been put up by the Raytheon, has 15 reference stations scattered across the country. Two mission control centres, along with associated uplink stations, have been set up at Kundalahalli in Bangalore. One more control centre and uplink station are to come up at Delhi. As a part of the programme, a network of 18 total electron content (TEC) monitoring stations were installed at various locations in India to study and analyse the behaviour of the ionosphere over the Indian region.

GAGAN's TDS signal in space provides three-metre accuracy as against the requirement of 7.6 metres. Flight inspection of GAGAN signal is being carried out at Kozhikode, Hyderabad, Nagpur and Bangalore airports and the results have been satisfactory so far.

Study of Ionosphere

One essential component of the GAGAN project is the study of the ionospheric behaviour over the Indian region. This has been specially taken up in view of the uncertain nature of the behaviour of the ionosphere in the region. The study will lead to the optimisation of the algorithms for the ionospheric corrections in the region.

To study the ionospheric behaviour more effectively over entire Indian airspace, Indian universities and R&D labs, which are involved in the development of regional based ionotropic model for GAGAN, have suggested nine more TEC stations.

Technology Integration

GAGAN is now in operational phase and is compatible with other SBAS systems such as the wide-area augmentation system (WAAS), the European Geostationary Navigation Overlay Service (EGNOS) and the Multi-functional Satellite Augmentation System (MSAS) and will provide seamless air navigation service across regional boundaries. While the ground segment consists of eight reference stations and a master control center, which will have sub systems such as data communication network, SBAS correction and verification system, operations and maintenance system, performance monitoring display and payload simulator, Indian land uplinking stations will have dish antenna assembly. The space segment will consist of one geo-navigation transponder.

Effective Flight-management System

A flight-management system based on GAGAN will then be poised to save operators time and money by managing climb, descent and engine performance profiles. The FMS will improve the efficiency and flexibility by increasing the use of operator-preferred trajectories. It will improve airport and airspace access in all weather conditions, and the ability to meet the environmental and obstacle clearance constraints. It will also enhance reliability and reduce delays by defining more precise terminal area procedures that feature parallel routes and environmentally optimised airspace corridors.

- GAGAN will increase safety by using a three-dimensional approach operation with course guidance to the runway, which will reduce the risk of controlled flight into terrain i.e., an accident whereby an airworthy aircraft, under pilot control, inadvertently flies into terrain, an obstacle, or water.
- GAGAN will also offer high position accuracies over a wide geographical area like the Indian airspace. These positions accuracies will be simultaneously available to 80 civilian and more than 200 non-civilian airports and airfields and will facilitate an increase in the number of airports to 500 as planned. These position accuracies can be further enhanced with ground-based, augmentation system.

Developments

The first GAGAN transmitter was integrated into the GSAT-4 geostationary satellite, and had a goal of being operational in 2008. Following a series of delays, GSAT-4 was launched on 15 April 2010, however it failed to reach orbit after the third stage of the Geosynchronous Satellite Launch Vehicle Mk.II that was carrying it malfunctioned.

In 2009, Raytheon had won an 82 million dollar contract. It was mainly dedicated to modernise Indian air navigation system. The vice president of Command and Control Systems, Raytheon Network Centric Systems, Andy Zogg commented:

> "GAGAN will be the world's most advanced air navigation system and further reinforces India's leadership in the forefront of air navigation. GAGAN will greatly improve safety, reduce congestion and enhance communications to meet India's growing air traffic management needs".

In 2012, the Defence Research and Development Organisation received a "miniaturised version" of the device with all the features from global positioning systems (GPS) and global navigation satellite systems (GNSS). The module weighing just 17 gm, can be used in multiple platforms ranging from aircraft (e.g. winged or rotor-craft) to small boats, ships. Reportedly, it can also assist "survey applications". It is a cost-efficient device and can be of "tremendous" civilian use. The navigation output is composed of GPS, GLONASS and GPS+GLONASS position, speed and time data. According to a statement released by the DRDO, G30M is a state-of-the-art technology receiver, integrating Indian GAGAN as well as both global positioning system and GLONASS systems.

On 30 December 2012, the Directorate General of Civil Aviation (DGCA), India provisionally certified the GPS-aided geo-augmented navigation (GAGAN) system to RNP0.1 (Required Navigation Performance, 0.1 Nautical Mile) service level. The certification enabled aircraft fitted with SBAS equipment to use GAGAN signal in space for navigation purposes.

Satellites

- GSAT-8 is an Indian geostationary satellites, which was successfully launched using Ariane 5 on 21 May 2011 and is positioned in geosynchronous orbit at 55 degrees E longitude.
- GSAT-10 is envisaged to augment the growing need of Ku and C-band transponders and carries 12 Ku Band, 12 C Band and 12 Extended C Band transponders and a GAGAN payload. The spacecraft employs the standard I-3K structure with power handling capability of around 6 kW with a lift off mass of 3400 kg. GSAT-10 was successfully launched by Ariane 5 on 29 September 2012.

- GSAT-15 carries 24 Ku band transponders with India coverage beam and a GAGAN payload was successfully launched on 10 November 2015, 21:34:07 UTC, completing the constellation.

Indian Regional Navigation Satellite System

The Indian government has stated that it intends to use the experience of creating the GAGAN system to enable the creation of an autonomous regional navigation system called the Indian Regional Navigation Satellite System IRNSS.

IRNSS-1 Indian regional navigational satellite system (IRNSS)-1, the first of the seven satellites of the IRNSS constellation, carries a navigation payload and a C-band ranging transponder. The spacecraft employs an optimised I-1K structure with a power handling capability of around 1660W and a lift off mass of 1425 kg, and is designed for a nominal mission life of 10 years. The first satellite of IRNSS constellation was launched onboard PSLV (C22) on 1 July 2012. While the full constellation was planned to be realised during 2014 time frame, launch of subsequent satellites got delayed.

Currently all 7 satellites are in orbit but in 2017 it was announced that all three rubidium atomic clocks on board IRNSS-1A had failed, mirroring similar failures in the Galileo constellation. The first failure occurred in July 2016, following which two other clocks also failed. This rendered the satellite somewhat redundant and required replacement. Although the satellite still performs other functions, the data is coarse, and thus cannot be used for accurate measurements. ISRO plans to replace it with IRNSS-1H in July or August 2017.

Two more clocks in the navigational system had started showing signs of abnormality, thereby taking the total number of failed clocks to five.

As a precaution to extend the operational life of navigation satellite, ISRO is running only one rubidium atomic clock instead of two in the remaining six satellites. Each satellite has three clocks, therefore a total of 27 clocks for all satellites in the system (including standby satellites). The clocks of both IRNSS and GALILEO were supplied by SpectraTime. ISRO replaced the atomic clocks in two standby NavIC satellites. The setback comes at a time when IRNSS is yet to start commercial operations.

Applications

Karnataka Forest Department has used GAGAN to build a new, accurate and publicly available satellite based database of its forestlands. This is a follow-up to the Supreme Court directive to states to update and put up their respective forest maps. The geospatial database of forestlands pilot has used data from the Cartosat-2 satellite. The maps are meant to rid authorities of ambiguities related to forest boundaries and give clarity to forest administrators, revenue officials as also the public.

The Indian National Centre for Ocean Information Services (INCOIS) along with AAI has launched a new satellite-based GEMINI (Gagan Enabled Mariner's Instrument for Navigation and Information) system that will alert deep-sea fishermen of upcoming disasters. The GEMINI app on the cellphone decodes the signals from GEMINI device and alerts the user on imminent threats like cyclones, high waves, strong winds along with PFZ and search and rescue mission.

Various Indian manufactured missiles including the BrahMos will use GAGAN for guidance.

Quasi-Zenith Satellite System

The Quasi-Zenith Satellite System (QZSS), also known as Michibiki, is a four-satellite regional time transfer system and a satellite-based augmentation system development by the Japanese government to enhance the United States-operated Global Positioning System (GPS) in the Asia-Oceania regions, with a focus on Japan. The goal of QZSS is to provide highly precise and stable positioning services in the Asia-Oceania region, compatible with GPS. Four-satellite QZSS services were available on a trial basis as of January 12, 2018, and officially started on November 1, 2018. A satellite navigation system independent of GPS is planned for 2023 with 7 satellites.

Quasi-Zenith satellite orbit.

Orbit

QZSS uses one geostationary satellite and three satellites in Tundra-type highly inclined, slightly elliptical, geosynchronous orbits. Each orbit is 120° apart from the other two. Because of this inclination, they are not geostationary; they do not remain in the same place in the sky. Instead, their ground traces are asymmetrical figure patterns (analemmas), designed to ensure that one is almost directly overhead (elevation 60° or more) over Japan at all times.

The nominal orbital elements are:

QZSS Satellite Keplerian Elements (Nominal)	
Epoch	2009-12-26 12:00 UTC
Semimajor axis (a)	42,164 kilometres (26,199 mi)
Eccentricity (e)	0.075 ± 0.015
Inclination (i)	43° ± 4°
Right ascension of the ascending node (Ω)	195° (initial)
Argument of perigee (ω)	270° ± 2°
Mean anomaly (M_0)	305° (initial)
Central longitude of ground trace	135° E ± 5°

Satellites

Current 4 Satellite Constellation:

Name	Launch date	Status
QZS-1 (Michibiki-1)	11 September 2010	Operational
QZS-2 (Michibiki-2)	1 June 2017	Operational
QZS-3 (Michibiki-3)	19 August 2017	Operational
QZS-4 (Michibiki-4)	10 October 2017	Operational

Future 7 Satellite Constellation:

Name	Planned Launch date	Status
QZS-1R	2020	Future
QZS-5	2022	Future
QZS-6	2023	Future
QZS-7	2023	Future

QZSS and Positioning Augmentation:

The primary purpose of QZSS is to increase the availability of GPS in Japan's numerous urban canyons, where only satellites at very high elevation can be seen. A secondary function is performance enhancement, increasing the accuracy and reliability of GPS derived navigation solutions.

The Quasi-Zenith Satellites transmit signals compatible with the GPS L1C/A signal, as well as the modernized GPS L1C, L2C signal and L5 signals. This minimizes changes to existing GPS receivers.

Compared to standalone GPS, the combined system GPS plus QZSS delivers improved positioning performance via ranging correction data provided through the transmission of submeter-class performance enhancement signals L1-SAIF and LEX from

QZSS. It also improves reliability by means of failure monitoring and system health data notifications. QZSS also provides other support data to users to improve GPS satellite acquisition.

According to its original plan, QZSS was to carry two types of space-borne atomic clocks; a hydrogen maser and a rubidium (Rb) atomic clock. The development of a passive hydrogen maser for QZSS was abandoned in 2006. The positioning signal will be generated by a Rb clock and an architecture similar to the GPS timekeeping system will be employed. QZSS will also be able to use a Two-Way Satellite Time and Frequency Transfer (TWSTFT) scheme, which will be employed to gain some fundamental knowledge of satellite atomic standard behavior in space as well as for other research purposes.

QZSS Timekeeping and Remote Synchronization

Although the first generation QZSS timekeeping system (TKS) will be based on the Rb clock, the first QZSS satellites will carry a basic prototype of an experimental crystal clock synchronization system. During the first half of the two year in-orbit test phase, preliminary tests will investigate the feasibility of the atomic clock-less technology which might be employed in the second generation QZSS.

The mentioned QZSS TKS technology is a novel satellite timekeeping system which does not require on-board atomic clocks as used by existing navigation satellite systems such as GPS, GLONASS, NAVIC or Galileo system. This concept is differentiated by the employment of a synchronization framework combined with lightweight steerable on-board clocks which act as transponders re-broadcasting the precise time remotely provided by the time synchronization network located on the ground. This allows the system to operate optimally when satellites are in direct contact with the ground station, making it suitable for a system like the Japanese QZSS. Low satellite mass and low satellite manufacturing and launch cost are significant advantages of this system. An outline of this concept as well as two possible implementations of the time synchronization network for QZSS were studied and published in Remote Synchronization Method for the Quasi-Zenith Satellite System and Remote Synchronization Method for the Quasi-Zenith Satellite System: study of a novel satellite timekeeping system which does not require on-board atomic clocks.

Local-area Augmentation Systems

The local-area augmentation system (LAAS) is an all-weather aircraft landing system based on real-time differential correction of the GPS signal. Local reference receivers located around the airport send data to a central location at the airport. This data is used to formulate a correction message, which is then transmitted to users via a VHF

Data Link. A receiver on an aircraft uses this information to correct GPS signals, which then provides a standard ILS-style display to use while flying a precision approach. The FAA has stopped using the term LAAS and has transitioned to the International Civil Aviation Organization (ICAO) terminology of Ground-Based Augmentation System (GBAS). While the FAA has indefinitely delayed plans for federal GBAS acquisition, the system can be purchased by airports and installed as a Non-Federal navigation aid.

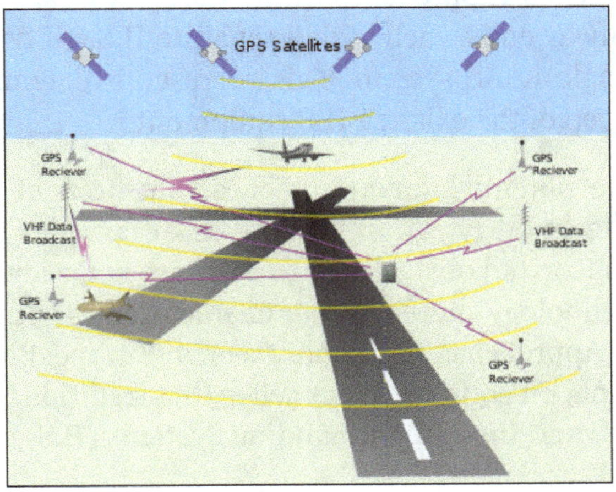

Operation

Local reference receivers are located around an airport at precisely surveyed locations. The signal received from the GPS constellation is used to calculate the position of the LAAS ground station, which is then compared to its precisely surveyed position. This data is used to formulate a correction message which is transmitted to users via a VHF data link. A receiver on the aircraft uses this information to correct the GPS signals it receives. This information is used to create an ILS-type display for aircraft approach and landing purposes. Honeywell's CAT I system provides precision approach service within a radius of 23 NM surrounding a single airport. LAAS mitigates GPS threats in the Local Area to a much greater accuracy than WAAS and therefore provides a higher level of service not attainable by WAAS. LAAS's VHF uplink signal is currently slated to share the frequency band from 108 MHz to 118 MHz with existing ILS localizer and VOR navigational aids. LAAS utilizes a Time Division Multiple Access (TDMA) technology in servicing the entire airport with a single frequency allocation. With future replacement of ILS, LAAS will reduce the congested VHF NAV band.

Accuracy

The current Category-1 (GAST-C) GBAS achieves a Category I Precision Approach accuracy of 16 m laterally and 4 m vertically. The goal of a to-be developed GAST-D GBAS is to provide Category III Precision Approach capability. The minimum accuracy for lateral and vertical errors of a Category III system are specified in RTCA DO-245A, Minimum Aviation System Performance Standards for Local Area Augmentation

System (LAAS). The GAST-D GBAS will allow aircraft to land with zero visibility using 'autoland' systems.

Benefits

One of the primary benefits of LAAS is that a single installation at a major airport can be used for multiple precision approaches within the local area. For example, if Chicago O'Hare has twelve runway ends, each with a separate ILS, all twelve ILS facilities can be replaced with a single LAAS system. This represents a significant cost savings in maintenance and upkeep of the existing ILS equipment.

Another benefit is the potential for approaches that are not straight- in. Aircraft equipped with LAAS technology can utilize curved or complex approaches such that they could be flown on to avoid obstacles or to decrease noise levels in areas surrounding an airport. This technology shares similar characteristics with the older Microwave landing system (MLS) Approaches, commonly seen in Europe. Both systems allow lower visibility requirements on complex approaches that traditional Wide Area Augmentation Systems (WAAS) and Instrument Landing Systems (ILS) could not allow.

The FAA also contends that only a single set of navigational equipment will be needed on an aircraft for both LAAS and WAAS capability. This lowers initial cost and maintenance per aircraft since only one receiver is required instead of multiple receivers for NDB's, DME, VOR, ILS, MLS and GPS. The FAA hopes this will result in decreased cost to the airlines and passengers as well as general aviation.

Drawbacks

LAAS shares the drawbacks of all RF based landing systems; those being jamming both intentional and accidental.

Variations

The Joint Precision Approach and Landing System (JPALS) is a similar system for military usage. Honeywell has developed the Honeywell International Satellite Landing System (SLS) 4000 series (SLS-4000) which received system design approval from the FAA on September 3, 2009, with a follow-on approval of an enhanced SLS-4000 (SLS-4000 Block 1) in September 2012.

Wide Area Augmentation System

The Wide Area Augmentation System (WAAS) provides extremely accurate navigation capability by augmenting the Global Positioning System (GPS). It was developed for civil aviation by the Federal Aviation Administration (FAA) and covers most of the U.S. National Airspace System (NAS) as well as parts of Canada and Mexico. WAAS provides service for all classes of aircraft in all phases of flight including en-route navigation, airport departures, and airport arrivals. This includes vertically guided landing approaches that can be used in Instrument Meteorological Conditions (IMC).

The Wide Area Augmentation System covers nearly all of the U.S. National Airspace System (NAS). The WAAS provides augmentation information to GPS receivers and thus enhances the accuracy and the reliability of position estimates. This is accomplished as follows:

- The signals from GPS satellites are received across the NAS at numerous widely-spaced Wide Area Reference Stations (WRS) sites. The location of these WRS sites are precisely surveyed to allow any errors in the received GPS signals to be detected.
- The GPS information collected by the WRS sites is transmitted to the WAAS Master Station (WMS) by means of a land-based communications network. Based on the information received, the WMS generates WAAS augmentation messages.
- The augmentation messages are sent from the WMS to uplink stations from where they are transmitted to the navigation payloads on Geostationary communications satellites.
- The Geostationary satellites then broadcast the augmentation messages on a GPS-like signal. These messages provide information that allows aircraft GPS/WAAS receivers to remove errors in the GPS signal, allowing for a significant increase in location accuracy and reliability.
- WAAS also provides indications to GPS/WAAS receivers of where the GPS system is unusable due to system errors or other issues.

Benefits

The Wide Area Augmentation System provides many benefits to aviation and also supports a number of non-aviation uses. Some of the primary benefits to aviation users are as follows:

- Safety: The WAAS system was designed to the strictest of safety standards users are notified within six seconds of any Global Positioning System issuance of hazardously misleading information that would cause an error in the GPS position estimate.
- Improved position accuracy: WAAS results in a significant improvement in both horizontal and vertical position accuracy.

- Improved airport access: Where authorized, WAAS capability can allow a RNAV (GPS) approach to be flown to LPV (Localizer Performance with Vertical guidance) minimums of as low as 200' which is equivalent to ILS Cat 1 capabilities. The number of WAAS enabled LPV approaches now far exceeds the number of ILS installations within the U.S. NAS.
- RNP 0.3 availability: WAAS can support Required Navigation Performance (RNP) values of as low as 0.3 in both North and South America.
- Interoperability: Satellite Based Augmentation Systems (SBAS), similar to WAAS, are available in Europe, Japan, India, and are being developed in other regions of the world. All of these systems are designed to be interoperable which allows aircraft equipped with WAAS receivers to use these other systems as well.

References

- GNSS-Augmentation: gssc.esa.int, Retrieved 19 June, 2020
- Fabrizio Tappero (2009-05-24). Remote Synchronization Method for the Quasi-Zenith Satellite System: study of a novel satellite timekeeping system which does not require on-board atomic clocks. VDM Verlag. ISBN 978-3-639-16004-8
- SBAS-Fundamentals: gssc.esa.int, Retrieved 25 April, 2020
- Wide-Area-Augmentation-System-(WAAS): skybrary.aero, Retrieved 28 August, 2020

3

Global Positioning System

Global Positioning System is a type of global navigation satellite system that gives geolocation and time information to a GPS receiver anywhere on Earth. It consists of a network of 24 orbiting satellites. This chapter has been carefully written to provide an easy understanding of all the aspects of global navigation system.

The Global Positioning System (GPS) is currently the only fully functional Global Navigation Satellite System (GNSS). More than two dozen GPS satellites are in medium Earth orbit, transmitting signals allowing GPS receivers to determine the receiver's location, speed and direction.

Since the first experimental satellite was launched in 1978, GPS has become an indispensable aid to navigation around the world, and an important tool for map-making and land surveying. GPS also provides a precise time reference used in many applications including scientific study of earthquakes, and synchronization of telecommunications networks.

Developed by the United States Department of Defense, it is officially named NAVSTAR GPS (NAVigation Satellite Timing and Ranging Global Positioning System). The satellite constellation is managed by the United States Air Force 50th Space Wing. The cost of maintaining the system is approximately US$750 million per year, including the replacement of aging satellites, and research and development. Despite this fact, GPS is free for civilian use as a public good.

Simplified Method of Operation

A GPS receiver calculates its position by measuring the distance between itself and three or more GPS satellites. Measuring the time delay between transmission and reception of each GPS radio signal gives the distance to each satellite, since the signal travels at a known speed. The signals also carry information about the satellites' location. By determining the position of, and distance to, at least three satellites, the receiver can compute its position using trilateration, or the determination of a distance from three points. Receivers typically do not have perfectly accurate clocks and therefore track one or more additional satellites to correct the receiver's clock error.

Technical Description

System Segmentation

The current GPS consists of three major segments: the space segment (SS), a control segment (CS), and a user segment (US).

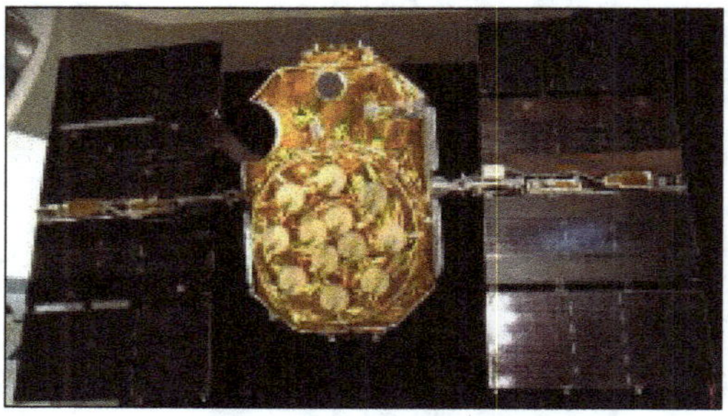

GPS satellite on test rack.

Space Segment

The SS is composed of the orbiting GPS satellites or Space Vehicles (SV) in GPS parlance. The GPS design calls for 24 SVs to be distributed equally among six circular orbital planes. The orbital planes are centered on the Earth, not rotating with respect to the distant stars. The six planes have approximately 55° inclination (tilt relative to Earth's equator) and are separated by 60° right ascension of the ascending node (angle along the equator from a reference point to the orbit's intersection).

Orbiting at an altitude of approximately 20,200 kilometers (12,600 miles or 10,900 nautical miles), each SV makes two complete orbits each sidereal day (the length of time for Earth to make a full rotation with respect to a fixed star, namely, 23 hours, 56 minutes, and 4.1 seconds), so it passes over the same location on Earth once each day. The orbits are arranged so that at least six satellites are always within line of sight from almost anywhere on Earth.

As of February 2007, there are 30 actively broadcasting satellites in the GPS constellation. The additional satellites improve the precision of GPS receiver calculations by providing redundant measurements. With the increased number of satellites, the constellation was changed to a nonuniform arrangement. Such an arrangement was shown to improve reliability and availability of the system, relative to a uniform system, when multiple satellites fail.

Control Segment

The flight paths of the satellites are tracked by US Air Force monitoring stations in Hawaii, Kwajalein, Ascension Island, Diego Garcia, and Colorado Springs, Colorado, along with monitor stations operated by the National Geospatial-Intelligence Agency (NGA). The tracking information is sent to the Air Force Space Command's master control station at Schriever Air Force Base, Colorado Springs, Colorado, which is operated by the 2d Space Operations Squadron (2 SOPS) of the United States Air Force (USAF). 2 SOPS contacts each GPS satellite regularly with a navigational update (using the ground antennas at Ascension Island, Diego Garcia, Kwajalein, and Colorado Springs). These updates synchronize the atomic clocks on board the satellites to within one microsecond and adjust the ephemeris of each satellite's internal orbital model. The updates are created by a Kalman Filter which uses inputs from the ground monitoring stations, space weather information, and other various inputs.

GPS receivers come in a variety of formats, from devices integrated into cars, phones, and watches.

User Segment

The user's GPS receiver is the user segment (U.S.) of the GPS system. In general, GPS receivers are composed of an antenna, tuned to the frequencies transmitted by the satellites, receiver-processors, and a highly-stable clock (often a crystal oscillator). They may also include a display for providing location and speed information to the user. A receiver is often described by its number of channels: This signifies how many satellites it can monitor simultaneously. Originally limited to four or five, this has progressively increased over the years so that, as of 2006, receivers typically have between twelve and twenty channels.

A typical OEM GPS receiver module, based on the SiRF Star III chipset, measuring 12 x 15mm, and used in many products.

Many GPS receivers can relay position data to a PC or other device using the NMEA 0183 protocol. NMEA 2000 is a newer and less widely adopted protocol. Both are proprietary and controlled by the US-based National Marine Electronics Association. References to the NMEA protocols have been compiled from public records, allowing open source tools like gpsd to read the protocol without violating intellectual property laws.

Other proprietary protocols exist as well, such as by SiRF Technology Inc. Receivers can interface with other devices using methods including a serial connection, USB or Bluetooth.

Navigation Signals

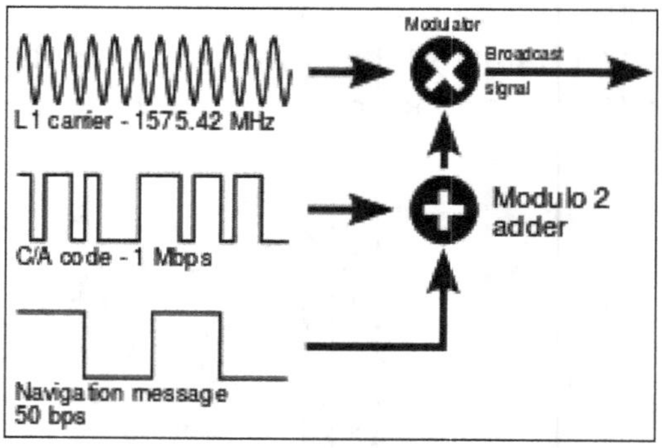

GPS broadcast signal.

GPS satellites broadcast three different types of data in the primary navigation signal. The first is the almanac which sends coarse time information along with status information about the satellites. The second is the ephemeris, which contains orbital information that allows the receiver to calculate the position of the satellite. This data is included in the 37,500 bit Navigation Message, which takes 12.5 minutes to send at 50 bps.

The satellites also broadcast two forms of clock information, the Coarse/Acquisition code, or C/A which is freely available to the public, and the restricted Precise code, or P-code, usually reserved for military applications. The C/A code is a 1,023 bit long pseudo-random code broadcast at 1.023 MHz, repeating every millisecond. Each satellite sends a distinct C/A code, which allows it to be uniquely identified. The P-code is a similar code broadcast at 10.23 MHz, but it repeats only once a week. In normal operation, the so-called "anti-spoofing mode" (spoofing, in GPS means a fake signal), the P code is first encrypted into the Y-code, or P(Y), and then decrypted by units with a valid decryption key. Frequencies used by GPS include:

- L1 (1575.42 MHz): Mix of Navigation Message, coarse-acquisition (C/A) code and encrypted precision P(Y) code.
- L2 (1227.60 MHz): P(Y) code, plus the new L2C code on the Block IIR-M and newer satellites.
- L3 (1381.05 MHz): Used by the Defense Support Program to signal detection of missile launches, nuclear detonations, and other high-energy infrared events.
- L4 (1379.913 MHz): Being studied for additional ionospheric correction.
- L5 (1176.45 MHz): Proposed for use as a civilian safety-of-life (SoL) signal. This

frequency falls into an internationally protected range for aeronautical navigation, promising little or no interference under all circumstances. The first Block IIF satellite that would provide this signal is set to be launched in 2008.

Calculating Positions

The coordinates are calculated according to the World Geodetic System WGS84 coordinate system. To calculate its position, a receiver needs to know the precise time. The satellites are equipped with extremely accurate atomic clocks, and the receiver uses an internal crystal oscillator-based clock that is continually updated using the signals from the satellites.

The receiver identifies each satellite's signal by its distinct C/A code pattern, then measures the time delay for each satellite. To do this, the receiver produces an identical C/A sequence using the same seed number as the satellite. By lining up the two sequences, the receiver can measure the delay and calculate the distance to the satellite, called the pseudorange.

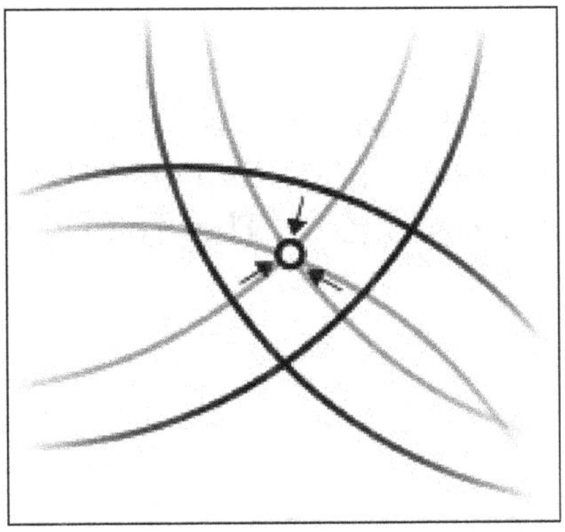

Overlapping pseudoranges, represented as curves, are modified to yield the probable position.

The orbital position data from the Navigation Message is then used to calculate the satellite's precise position. Knowing the position and the distance of a satellite indicates that the receiver is located somewhere on the surface of an imaginary sphere centered on that satellite and whose radius is the distance to it. When four satellites are measured simultaneously, the intersection of the four imaginary spheres reveals the location of the receiver. Earth-based users can substitute the sphere of the planet for one satellite by using their altitude. Often, these spheres will overlap slightly instead of meeting at one point, so the receiver will yield a mathematically most-probable position (and often indicate the uncertainty).

Calculating a position with the P(Y) signal is generally similar in concept, assuming one can decrypt it. The encryption is essentially a safety mechanism; if a signal can be successfully decrypted, it is reasonable to assume it is a real signal being sent by a GPS satellite. In comparison, civil receivers are highly vulnerable to spoofing since correctly formatted C/A signals can be generated using readily available signal generators.

RAIM features will not help, since RAIM only checks the signals from a navigational perspective.

Accuracy and Error Sources

The position calculated by a GPS receiver requires the current time, the position of the satellite and the measured delay of the received signal. The position accuracy is primarily dependent on the satellite position and signal delay.

To measure the delay, the receiver compares the bit sequence received from the satellite with an internally generated version. By comparing the rising and trailing edges of the bit transitions, modern electronics can measure signal offset to within about 1 percent of a bit time, or approximately 10 nanoseconds for the C/A code. Since GPS signals propagate nearly at the speed of light, this represents an error of about 3 meters. This is the minimum error possible using the GPS C/A signal.

Position accuracy can be improved by using the higher-speed P(Y) signal. Assuming the same 1 percent accuracy, the faster P(Y) signal results in accuracy of about 30 centimeters.

Electronics errors are one of several accuracy-degrading effects outlined in the table below. When taken together, autonomous civilian GPS horizontal position fixes are typically accurate to about 15 meters (50 ft). These effects also reduce the more precise P(Y) code's accuracy.

Sources of Errors	
Source	Effect
Ionospheric effects	± 5 meter
Ephemeris errors	± 2.5 meter
Satellite clock errors	± 2 meter
Multipath distortion	± 1 meter
Tropospheric effects	± 0.5 meter
Numerical errors	± 1 meter or less

Atmospheric Effects

Changing atmospheric conditions change the speed of the GPS signals as they pass through the Earth's atmosphere and ionosphere. Correcting these errors is a significant challenge to improving GPS position accuracy. These effects are minimized when the satellite is directly overhead, and become greater for satellites nearer the horizon, since the signal is affected for a longer time. Once the receiver's approximate location is known, a mathematical model can be used to estimate and compensate for these errors.

Because ionospheric delay affects the speed of radio waves differently based on frequency, a characteristic known as dispersion (a phenomenon where a light signal is separated into its constituent elements), both frequency bands can be used to help reduce this error. Some military and expensive survey-grade civilian receivers compare the different delays in the L1 and L2 frequencies to measure atmospheric dispersion, and apply a more precise correction. This can be done in civilian receivers without decrypting the P(Y) signal carried on L2, by tracking the carrier wave instead of the modulated code. To facilitate this on lower cost receivers, a new civilian code signal on L2, called L2C, was added to the Block IIR-M satellites, first launched in 2005. It allows a direct comparison of the L1 and L2 signals using the coded signal instead of the carrier wave.

The effects of the ionosphere are generally slow-moving, and can be averaged over time. The effects for any particular geographical area can be easily calculated by comparing the GPS-measured position to a known surveyed location. This correction is also valid for other receivers in the same general location. Several systems send this information over radio or other links to allow only L1 receivers to make ionospheric corrections. The ionospheric data are transmitted via satellite in Satellite Based Augmentation Systems such as Wide Area Augmentation System (WAAS), which transmits it on the GPS frequency using a special pseudo-random number (PRN), so only one antenna and receiver are required.

Humidity also causes a variable delay, resulting in errors similar to ionospheric delay, but occurring in the troposphere. This effect is much more localized, and changes more quickly than the ionospheric effects, making precise compensation for humidity more difficult. Altitude also causes a variable delay, as the signal passes through fewer atmospheres at higher elevations. Since the GPS receiver measures altitude directly, this is a much simpler correction to apply.

Multipath Effects

GPS signals can also be affected by multipath issues, where the radio signals reflect off surrounding terrain such as buildings, canyon walls, and hard ground. These delayed signals can cause inaccuracy. A variety of techniques, most notably narrow correlator spacing, have been developed to mitigate multipath errors. For long delay multipath, the receiver itself can recognize the wayward signal and discard it. To address shorter delay multipath from the signal reflecting off the ground, specialized antennas may be used. Short delay reflections are harder to filter out since they are only slightly delayed, causing effects almost indistinguishable from routine fluctuations in atmospheric delay.

Multipath effects are much less severe in moving vehicles. When the GPS antenna is moving, the false solutions using reflected signals quickly fail to converge and only the direct signals result in stable solutions.

Ephemeris and Clock Errors

The navigation message from a satellite is sent out only every 12.5 minutes. In reality, the data contained in these messages tend to be "out of date" by a greater distance.

Consider the case when a GPS satellite is boosted back into a proper orbit; for some time following the maneuver, the receiver's calculation of the satellite's position will be incorrect until it receives another ephemeris update. The onboard clocks are extremely accurate, but they do suffer from some clock drift. This problem tends to be very small, but may add up to 2 meters (~6 ft.) of inaccuracy.

This class of error is more "stable" than ionospheric problems and tends to change over days or weeks rather than minutes. This makes correction fairly simple by sending out a more accurate almanac on a separate channel.

Selective Availability

The GPS includes a feature called Selective Availability (SA) that introduces intentional errors between 0 meters and up to a hundred meters (300 ft.) into the publicly available navigation signals, making it difficult to use for guiding long range missiles to precise targets. Additional accuracy was available in the signal, but in an encrypted form that was only available to the United States military, its allies, and a few others, mostly government users.

Relativity

According to the theory of relativity, due to their constant movement and height relative to the Earth-centered inertial reference frame, the clocks on the satellites are affected by their speed (special relativity) as well as their gravitational potential (general relativity). For the GPS satellites, general relativity predicts that the atomic clocks at GPS orbital altitudes will tick more rapidly, by about 45,900 nanoseconds (ns) per day, because they are in a weaker gravitational field than atomic clocks on Earth's surface. Special relativity predicts that atomic clocks moving at GPS orbital speeds will tick more slowly, by about 7,200 ns per day, than stationary ground clocks. When combined, the discrepancy is 38 microseconds per day; a difference of 4.465 parts in 1010. To account for this, the frequency standard onboard each satellite is given a rate offset prior to launch, making it run slightly slower than the desired frequency on Earth; specifically, at 10.22999999543 MHz instead of 10.23 MHz.

Another relativistic effect to be compensated for in GPS observation processing is the Sagnac effect. The GPS time scale is defined in an inertial system but observations are processed in an Earth-centered, Earth-fixed (co-rotating) system; a system in which simultaneity is not uniquely defined. The Lorentz transformation between the two systems modifies the signal run time, a correction having opposite algebraic signs for satellites in the Eastern and Western celestial hemispheres. Ignoring this effect will produce an East-West error on the order of hundreds of nanoseconds, or tens of meters in position.

The atomic clocks on board the GPS satellites are precisely tuned, making the system a practical engineering application of the scientific theory of relativity in a real-world system.

GPS Interference and Jamming

Since GPS signals at terrestrial receivers tend to be relatively weak, it is easy for other sources of electromagnetic radiation to overpower the receiver, making acquiring and tracking the satellite signals difficult or impossible. Solar flares are one such naturally occurring emission with the potential to degrade GPS reception, and their impact can affect reception over the half of the Earth facing the sun. GPS signals can also be interfered with by naturally occurring geomagnetic storms, predominantly found at near the poles of the Earth's magnetic field. Human-made interference can also disrupt, or jam, GPS signals. In one well-documented case, an entire harbor was unable to receive GPS signals due to unintentional jamming caused by a malfunctioning TV antenna preamplifier.

Intentional jamming is also possible. Generally, stronger signals can interfere with GPS receivers when they are within radio range, or line of sight. In 2002, a detailed description of how to build a short range GPS L1 C/A jammer was published in the online magazine Phrack. The U.S. government believes that such jammers were used occasionally during the Afghanistan War and the U.S. military claimed to destroy a GPS jammer with a GPS-guided bomb during the Iraq War. While a jammer is relatively easy to detect and locate, making it an attractive target for anti-radiation missiles, there is the possibility that these jammers would then be located near non-combatant infrastructure and used to attract the precision-guided munitions towards; a tactic known as using a human shield.

Due to the potential for both natural and man-made noise, numerous techniques continue to be developed to deal with the interference. The first is to not rely on GPS as a sole source. "IFR pilots should have a fallback plan in case of a GPS malfunction". Receiver Autonomous Integrity Monitoring (RAIM) is a feature now included in some receivers, which is designed to provide a warning to the user if jamming or another problem is detected. The military has also deployed the Selective Availability/Anti-Spoofing Module (SAASM) in its Defense Advanced GPS Receiver, which as shown in demonstration videos, is able to detect the jamming and maintain a lock on the GPS signals.

Techniques to Improve Accuracy

Augmentation

Augmentation methods of improving accuracy rely on external information being integrated into the calculation process. There are many such systems in place and they are generally named or described based on how the GPS sensor receives the information. Some systems transmit additional information about sources of error (such as clock drift, ephemeris, or ionospheric delay), others provide direct measurements of how much the signal was off in the past, while a third group provide additional navigational or vehicle information to be integrated in the calculation process.

Examples of augmentation systems include the WAAS, Differential GPS, and Inertial Navigation Systems.

Precise Monitoring

The accuracy of a calculation can also be improved through precise monitoring and measuring of the existing GPS signals in additional or alternate ways.

The first is called Dual Frequency monitoring, and refers to systems that can compare two or more signals, such as the L1 frequency to the L2 frequency. Since these are two different frequencies, they are affected in different, yet predictable ways by the atmosphere and objects around the receiver. After monitoring these signals, it is possible to calculate how much error is being introduced and then nullify that error.

Receivers that have the correct decryption key can relatively easily decode the P(Y)-code transmitted on both L1 and L2 to measure the error. Receivers that do not possess the key can still use a process called codeless to compare the encrypted information on L1 and L2 to gain much of the same error information. However, this technique is currently limited to specialized surveying equipment. In the future, additional civilian codes are expected to be transmitted on the L2 and L5 frequencies. When these become operational, non-encrypted users will be able to make the same comparison and directly measure some errors.

A second form of precise monitoring is called Carrier-Phase Enhancement (CPGPS). The error, which this corrects, arises because the pulse transition of the PRN is not instantaneous, and thus the correlation (satellite-receiver sequence matching) operation is imperfect. The CPGPS approach utilizes the L1 carrier wave, which has a period 1000 times smaller than that of the C/A bit period, to act as an additional clock signal and resolve the uncertainty. The phase difference error in the normal GPS amounts to between 2 and 3 meters (6 to 10 feet) of ambiguity. CPGPS working to within 1 percent of perfect transition reduces this error to 3 cm (1 inch) of ambiguity. By eliminating this source of error, CPGPS coupled with DGPS normally realizes between 20 and 30 cm (8 to 12 in) of absolute accuracy.

Relative Kinematic Positioning (RKP) is another approach for a precise GPS-based positioning system. In this approach, determination of range signal can be resolved to an accuracy of less than 10 cm (4 in). This is done by resolving the number of cycles in which the signal is transmitted and received by the receiver. This can be accomplished by using a combination of differential GPS (DGPS) correction data, transmitting GPS signal phase information and ambiguity resolution techniques via statistical tests—possibly with processing in real-time (real-time kinematic positioning, RTK).

GPS Time and Date

While most clocks are synchronized to Coordinated Universal Time (UTC), the Atomic clocks on the satellites are set to GPS time. The difference is that GPS time is not corrected to match the rotation of the Earth, so it does not contain leap seconds or other corrections which are periodically added to UTC. GPS time was set to match Coordinated Universal Time (UTC) in 1980, but has since diverged. The lack of corrections means that GPS time remains synchronized with the International Atomic Time (TAI).

The GPS navigation message includes the difference between GPS time and UTC, which as of 2006 is 14 seconds. Receivers subtract this offset from GPS time to calculate UTC and 'local' time. New GPS units may not show the correct UTC time until after receiving the UTC offset message. The GPS-UTC offset field can accommodate 255 leap seconds (eight bits) which, at the current rate of change of the Earth's rotation, is sufficient to last until the year 2330.

As opposed to the year, month, and day format of the Julian calendar, the GPS date is expressed as a week number and a day-of-week number. The week number is transmitted as a ten-bit field in the C/A and P(Y) navigation messages, and so it becomes zero again every 1,024 weeks (19.6 years). GPS week zero started at 00:00:00 UTC (00:00:19 TAI) on January 6 1980 and the week number became zero again for the first time at 23:59:47 UTC on August 21 1999 (00:00:19 TAI). In order to determine the current Gregorian date, a GPS receiver must be provided with the approximate date (to within 3,584 days) in order to correctly translate the GPS date signal. To address this concern the modernized GPS navigation messages use a 13-bit field, this repeats every 8,192 weeks (157 years), and will not return to zero until near the year 2137.

Applications

Military

GPS allows accurate targeting of various military weapons including cruise missiles and precision-guided munitions. To help prevent GPS guidance from being used in enemy or improvised weaponry, the US Government controls the export of civilian receivers. A US-based manufacturer cannot generally export a receiver unless the receiver contains limits restricting it from functioning when it is simultaneously at an altitude above 18 kilometers (60,000ft) and traveling at over 515 m/s (1,000 knots).

The GPS satellites also carry nuclear detonation detectors, which form a major portion of the United States Nuclear Detonation Detection System.

Navigation

- Automobiles can be equipped with GPS receivers at the factory or as after-market equipment. Units often display moving maps and information about location, speed, direction, and nearby streets and landmarks.
- Aircraft navigation systems usually display a "moving map" and are often connected to the autopilot for en-route navigation. Cockpit-mounted GPS

receivers and glass cockpits are appearing in general aviation aircraft of all sizes, using technologies such as WAAS or Local Area Augmentation System (LAAS) to increase accuracy. Many of these systems may be certified for instrument flight rules navigation, and some can also be used for final approach and landing operations. Glider pilots use GNSS Flight Recorders to log GPS data verifying their arrival at turn points in gliding competitions. Flight computers installed in many gliders also use GPS to compute wind speed aloft, and glide paths to waypoints such as alternate airports or mountain passes, to aid en route decision making for cross-country soaring.

- Boats and ships can use GPS to navigate all of the world's lakes, seas, and oceans. Maritime GPS units include functions useful on water, such as "man overboard" (MOB) functions that allow instantly marking the location where a person has fallen overboard, which simplifies rescue efforts. GPS may be connected to the ships self-steering gear and Chart plotters using the NMEA 0183 interface. GPS can also improve the security of shipping traffic by enabling AIS.
- Heavy Equipment can use GPS in construction, mining and precision agriculture. The blades and buckets of construction equipment are controlled automatically in GPS-based machine guidance systems. Agricultural equipment may use GPS to steer automatically, or as a visual aid displayed on a screen for the driver. This is very useful for controlled traffic and row crop operations and when spraying. Harvesters with yield monitors can also use GPS to create a yield map of the paddock being harvested.
- Bicycles often use GPS in racing and touring. GPS navigation allows cyclists to plot their course in advance and follow this course, which may include quieter, narrower streets, without having to stop frequently to refer to separate maps. Some GPS receivers are specifically adapted for cycling with special mounts and housings.
- Hikers, climbers, and even ordinary pedestrians in urban or rural environments can use GPS to determine their position, with or without reference to separate maps. In isolated areas, the ability of GPS to provide a precise position can greatly enhance the chances of rescue when climbers or hikers are disabled or lost (if they have a means of communication with rescue workers).
- GPS equipment for the visually impaired is available.
- Spacecraft are now beginning to use GPS as a navigational tool. The addition of a GPS receiver to a spacecraft allows precise orbit determination without ground tracking. This, in turn, enables autonomous spacecraft navigation, formation flying, and autonomous rendezvous. The use of GPS in MEO, GEO, HEO, and highly elliptical orbits is feasible only if the receiver can acquire and track the much weaker (15 20 dB) GPS side-lobe signals. This design constraint, and the radiation environment found in space, prevents the use of COTS receivers.

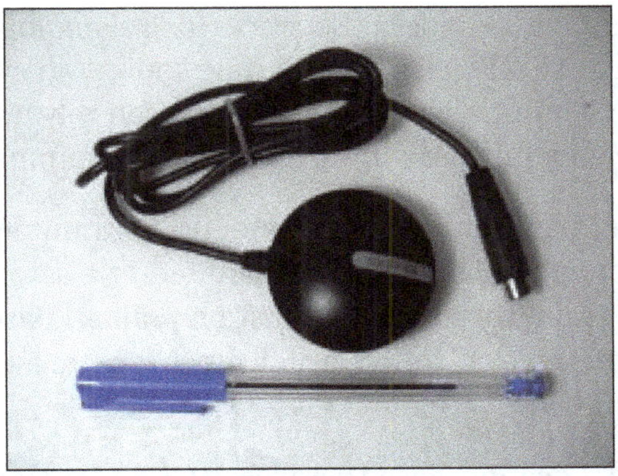

A modern SiRF Star III chip based 20-channel GPS receiver with WAAS/EGNOS support.

Surveying and Mapping

- Surveying: Survey-Grade GPS receivers can be used to position survey markers, buildings, and road construction. These units use the signal from both the L1 and L2 GPS frequencies. Even though the L2 code data are encrypted, the signal's carrier wave enables correction of some ionospheric errors. These dual-frequency GPS receivers typically cost US$10,000 or more, but can have positioning errors on the order of one centimeter or less when used in carrier phase differential GPS mode.
- Mapping and geographic information systems (GIS): Most mapping grade GPS receivers use the carrier wave data from only the L1 frequency, but have a precise crystal oscillator which reduces errors related to receiver clock jitter. This allows positioning errors on the order of one meter or less in real-time, with a differential GPS signal received using a separate radio receiver. By storing the carrier phase measurements and differentially post-processing the data, positioning errors on the order of 10 cm are possible with these receivers.
- Geophysics and geology: High precision measurements of crustal strain can be made with differential GPS by finding the relative displacement between GPS sensors. Multiple stations situated around an actively deforming area (such as a volcano or fault zone) can be used to find strain and ground movement. These measurements can then be used to interpret the cause of the deformation, such as a dike or sill beneath the surface of an active volcano.

Other Uses

This antenna is mounted on the roof of a hut containing a scientific experiment needing precise timing.

- Precise time reference: Many systems that must be accurately synchronized use GPS as a source of accurate time. GPS can be used as a reference clock for time code generators or NTP clocks. Sensors (for seismology or other monitoring application), can use GPS as a precise time source, so events may be timed accurately. TDMA communications networks often rely on this precise timing to synchronize RF generating equipment, network equipment, and multiplexers.

- Mobile Satellite Communications: Satellite communications systems use a directional antenna (usually a "dish") pointed at a satellite. The antenna on a moving ship or train, for example, must be pointed based on its current location. Modern antenna controllers usually incorporate a GPS receiver to provide this information.
- Emergency and Location-based services: GPS functionality can be used by emergency services to locate cell phones. The ability to locate a mobile phone is required in the United States by E911 emergency services legislation. However, as of September 2006 such a system is not in place in all parts of the country. GPS is less dependent on the telecommunications network topology than radiolocation for compatible phones. Assisted GPS reduces the power requirements of the mobile phone and increases the accuracy of the location. A phone's geographic location may also be used to provide location-based services including advertising, or other location-specific information.
- Location-based games: The availability of hand-held GPS receivers has led to games such as Geocaching, which involves using a hand-held GPS unit to travel to a specific longitude and latitude to search for objects hidden by other geocachers. This popular activity often includes walking or hiking to natural locations. Geodashing is an outdoor sport using waypoints.
- Aircraft passengers: Most airlines allow passenger use of GPS units on their flights, except during landing and take-off when other electronic devices are also restricted. Even though consumer GPS receivers have a minimal risk of interference, a few airlines disallow use of hand-held receivers during flight. Other airlines integrate aircraft tracking into the seat-back television entertainment system, available to all passengers even during takeoff and landing.
- Heading information: The GPS system can be used to determine heading information, even though it was not designed for this purpose. A "GPS compass" uses a pair of antennas separated by about 50 cm to detect the phase difference in the carrier signal from a particular GPS satellite. Given the positions of the satellite, the position of the antenna, and the phase difference, the orientation of the two antennas can be computed. More expensive GPS compass systems use three antennas in a triangle to get three separate readings with respect to each satellite. A GPS compass is not subject to magnetic declination as a magnetic compass is, and doesn't need to be reset periodically like a gyrocompass. It is, however, subject to multipath effects.

- GPS tracking systems use GPS to determine the location of a vehicle, person, or pet and to record the position at regular intervals in order to create a log of movements. The data can be stored inside the unit, or sent to a remote computer by radio or cellular modem. Some systems allow the location to be viewed in real-time on the Internet with a web-browser.
- Weather Prediction Improvements: Measurement of atmospheric bending of GPS satellite signals by specialized GPS receivers in orbital satellites can be used to determine atmospheric conditions such as air density, temperature, moisture and electron density. Such information from a set of six micro-satellites, launched in April 2006, called the Constellation of Observing System for Meteorology, Ionosphere and Climate COSMIC has been proven to improve the accuracy of weather prediction models.
- Photograph annotation: Combining GPS position data with photographs taken with a (typically digital) camera, allows one to lookup the locations where the photographs were taken in a gazeteer, and automatically annotate the photographs with the name of the location they depict. The GPS device can be integrated into the camera, or the timestamp of a picture's metadata can be combined with a GPS track log.
- Skydiving: Most commercial drop zones use a GPS to aid the pilot to "spot" the plane to the correct position relative to the dropzone that will allow all skydivers on the load to be able to fly their canopies back to the landing area. The "spot" takes into account the number of groups exiting the plane and the upper winds. In areas where skydiving through cloud is permitted the GPS can be the sole visual indicator when spotting in overcast conditions, this is referred to as a "GPS Spot".
- Marketing: Some market research companies have combined GIS systems and survey based research to help companies to decide where to open new branches, and to target their advertising according to the usage patterns of roads and the socio-demographic attributes of residential zones.

GPS Satellite Blocks

A GPS satellite is a satellite used by the NAVSTAR Global Positioning System (GPS). The first satellite in the system, Navstar 1, was launched February 22, 1978. The GPS satellite constellation is operated by the 50th Space Wing of the United States Space Force.

The GPS satellites circle the Earth at an altitude of about 20,000 km (12,427 miles) and complete two full orbits every day.

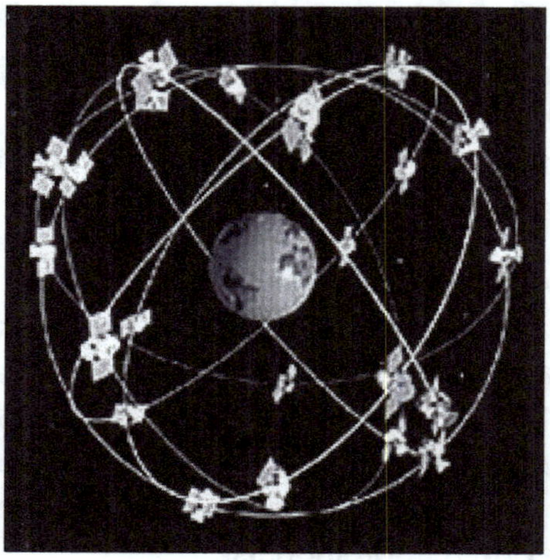

GPS satellite constellation.

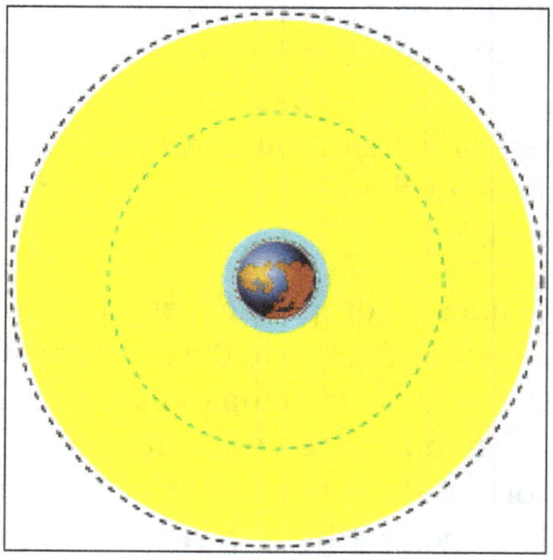

Various earth orbits to scale; green dash-dot line is medium earth orbit, a typical GPS orbit.

Block I Satellites

Rockwell International was awarded a contract in 1974 to build the first eight Block I satellites. In 1978 the contract was extended to build an additional three Block I satellites. Beginning with Navstar 1 in 1978, ten "Block I" GPS satellites were successfully launched. One satellite, "Navstar 7", was lost due to an unsuccessful launch on 18 December 1981.

The Block I satellites were launched from Vandenberg Air Force Base using Atlas rockets that were converted intercontinental ballistic missiles. The satellites were built by Rockwell International at the same plant in Seal Beach, CA where the S-II second stages of the Saturn V rockets had been built.

The Block I series consisted of the concept validation satellites and reflected various stages of system development. Lessons learned from the 11 satellites in the series were incorporated into the fully operational Block II series.

Dual solar arrays supplied over 400 watts of power, charging NiCd batteries for operations in Earth's shadow. S-band communications were used for control and telemetry, while a UHF channel provided cross-links between spacecraft. A hydrazine propulsion system was used for orbital correction. The payload included two L-band navigation signals at 1575.42 MHz (L1) and 1227.60 MHz (L2).

The final Block I launch was conducted on 9 October 1985, but the last Block I satellite was not taken out of service until 18 November 1995, well past its 5-year design life.

Block II Satellites

Initial Block II Series

The Block II satellites were the first full scale operational GPS satellites, designed to provide 14 days of operation without any contact from the control segment. The prime contractor was Rockwell International, which built a SVN 12 qualification vehicle after an amendment to the Block I contract. In 1983 the company was awarded an additional contract to build 28 Block II/IIA satellites.

Block II spacecraft were 3-axis stabilized, with ground pointing using reaction wheels. Two solar arrays supplied 710 watts of power, while S band communications were used for control and telemetry. A UHF channel was used for cross-links between spacecraft. A hydrazine propulsion system was used for orbital correction. The payload included two L band navigation signals at 1575.42 MHz (L1) and 1227.60 MHz (L2). Each spacecraft carried 2 rubidium and 2 cesium clocks, as well as nuclear detonation detection sensors, leading to a mass of 1,660 kilograms.

The first of the nine satellites in the initial Block II series was launched February 14, 1989; the last was launched October 1, 1990. The final satellite of the series to be taken out of service was decommissioned on March 15, 2007, well past its 7.5 year design life.

Block IIA Series

The Block IIA satellites were slightly improved versions of the Block II series, designed to provide 180 days of operation without contact from the control segment. However, the mass increased to 1,816 kilograms.

Nineteen satellites in the Block IIA series were launched, the first on November 26, 1990 and the last on November 6, 1997. The last satellite of the Block IIA (SVN-34), broadcast on the PRN 18 signal. It was removed from service October 9, 2019.

Two of the satellites in this series, numbers 35 and 36, are equipped with laser retro-reflectors, allowing them to be tracked independently of their radio signals, providing unambiguous separation of clock and ephemeris errors.

Block IIR Series

Artist's impression of a GPS-IIR satellite in orbit.

The Block IIR series are "replenishment" (replacement) satellites developed by Lockheed Martin. Each satellite weighs 4,480 pounds (2,030 kg) at launch and 2,370 pounds (1,080 kg) once on orbit. The first attempted launch of a Block IIR satellite failed on January 17, 1997 when the Delta II rocket exploded 12 seconds into flight. The first successful launch was on July 23, 1997. Twelve satellites in the series were successfully launched. At least ten satellites in this block carried an experimental S-band payload for search and rescue, known as Distress Alerting Satellite System.

Block IIR-M series

Artist's impression of a GPS-IIRM satellite in orbit.

The Block IIR-M satellites include a new military signal and a more robust civil signal, known as L2C. There are eight satellites in the Block IIR-M series, which were built by Lockheed Martin. The first Block IIR-M satellite was launched on September 26, 2005. The final launch of a IIR-M was on August 17, 2009.

Block IIF Series

Artist's impression of a Navstar-2F satellite in orbit.

The Block IIF series are "follow-on" satellites developed by Boeing. The satellite has a mass of 1,630 kilograms and a design life of 12 years. The first Block IIF space vehicle was launched in May 2010 on a Delta IV rocket. The twelfth and final IIF launch was February 5, 2016.

Block III Satellites

Block IIIA Series

Artist's impression of a GPS Block IIIA satellite in orbit.

GPS Block IIIA is the first series of third-generation GPS satellites, incorporating new signals and broadcasting at higher power levels. In September 2016, the Air Force awarded Lockheed Martin a contract option for two more Block IIIA satellites, setting the total number of GPS IIIA satellites to ten. On December 23, 2018, the first GPS III satellite was launched aboard a SpaceX Falcon 9. On August 22, 2019, the second GPS III satellite was launched aboard a Delta IV.

Block IIIF Series

The Block IIIF series is the second set of GPS Block III satellites, which will consist of up to 22 space vehicles. Block IIIF launches are expected to begin no earlier than 2025 and continue through 2034.

GPS Navigation Device

A Satellite navigation device, colloquially called a GPS receiver, or simply a GPS, is a device that is capable of receiving information from GNSS satellites and then to calculate the device's geographical position. Using suitable software, the device may display the position on a map, and it may offer routing directions. The Global Positioning System (GPS) is one of a handful of global navigation satellite systems (GNSS) made up of a network of a minimum of 24, but currently 30, satellites placed into orbit by the U.S. Department of Defense.

GPS was originally developed for use by the United States military, but in the 1980s, the United States government allowed the system to be used for civilian purposes. Though the GPS satellite data is free and works anywhere in the world, the GPS device and the associated software must be bought or rented.

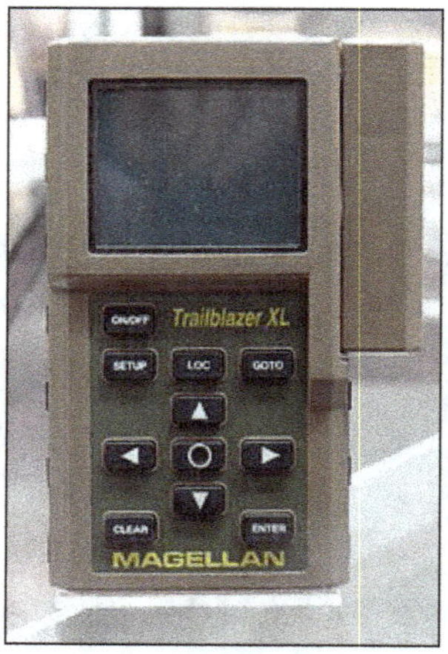

A 1993 GPS handheld receiver.

A satellite navigation device can retrieve (from one or more satellite systems) location and time information in all weather conditions, anywhere on or near the Earth. GPS reception requires an unobstructed line of sight to four or more GPS satellites, and is subject to poor satellite signal conditions. In exceptionally poor signal conditions, for example in urban areas, satellite signals may exhibit multipath propagation where signals bounce off structures, or are weakened by meteorological conditions. Obstructed lines of sight may arise from a tree canopy or inside a structure, such as in a building, garage or tunnel. Today, most standalone GPS receivers are used in automobiles. The GPS capability of smartphones may use assisted GPS (A-GPS) technology, which can use the base station or cell towers to provide a faster Time to First Fix (TTFF), especially when GPS signals are poor or unavailable. However, the mobile network part of the A-GPS technology would not be available when the smartphone is outside the range of the mobile reception network, while the GPS aspect would otherwise continue to be available.

Vehicle navigation on a personal navigation assistant.

The Russian Global Navigation Satellite System (GLONASS) was developed contemporaneously with GPS, but suffered from incomplete coverage of the globe until the mid-2000s.

GLONASS can be added to GPS devices to make more satellites available and enabling positions to be fixed more quickly and accurately, to within 2 meters.

Other satellite navigation services with (intended) global coverage are the European Galileo and the Chinese BeiDou.

Garmin eTrex10 edition handheld.

Automotive Navigation System

Using satellite information and subject to the sophistication of installed software, a GPS device used as an automobile navigation system may be used in a number of contexts, including:

- Maps, including street maps, displayed in human-readable format via text or in a graphical format.
- Turn-by-turn navigation directions to a human in charge of a vehicle or vessel via text or speech.
- Directions fed directly to an autonomous vehicle such as a robotic probe.
- Traffic congestion maps (depicting either historical or real-time data) and suggested alternative directions.
- Information on nearby amenities such as restaurants, fueling stations, and tourist attractions.

Navigation devices may be able to indicate:

- The roads or paths available.
- Traffic congestion and alternative routes.
- Roads or paths that might be taken to get to the destination.
- If some roads are busy (now or historically) the best route to take.

- The location of food, banks, hotels, fuel, airports or other places of interests.
- The shortest route between the two locations.
- The different options to drive on highway or back roads.

Sensitivity

GNSS devices vary in sensitivity, speed, vulnerability to multipath propagation, and other performance parameters. High Sensitivity receivers use large banks of correlators and digital signal processing to search for signals very quickly. This results in very fast times to first fix when the signals are at their normal levels, for example outdoors. When signals are weak, for example indoors, the extra processing power can be used to integrate weak signals to the point where they can be used to provide a position or timing solution.

GNSS signals are already very weak when they arrive at the Earth's surface. The GPS satellites only transmit 27 W (14.3 dBW) from a distance of 20,200 km in orbit above the Earth. By the time the signals arrive at the user's receiver, they are typically as weak as −160 dBW, equivalent to one-tenth of a million-billionth of a watt (100 attowatts). This is well below the thermal noise level in its bandwidth. Outdoors, GPS signals are typically around the −155 dBW level (−125 dBm).

Conventional GPS receivers integrate the received GPS signals for the same amount of time as the duration of a complete C/A code cycle which is 1 ms. This results in the ability to acquire and track signals down to around the −160 dBW level. High Sensitivity GPS receivers are able to integrate the incoming signals for up to 1,000 times longer than this and therefore acquire signals up to 1,000 times weaker, resulting in an integration gain of 30 dB. A good High Sensitivity GPS receiver can acquire signals down to −185 dBW, and tracking can be continued down to levels approaching −190 dBW.

High Sensitivity GPS can provide positioning in many but not all indoor locations. Signals are either heavily attenuated by the building materials or reflected as in multipath. Given that High Sensitivity GPS receivers may be up to 30 dB more sensitive, this is sufficient to track through 3 layers of dry bricks, or up to 20 cm (8 inches) of steel-reinforced concrete for example.

Examples of high sensitivity receiver chips include SiRFstarIII and MediaTek's MTK II.

Consumer Applications

Consumer GNSS navigation devices include:

- Dedicated GNSS navigation devices.
- Modules that need to be connected to a computer to be used.

- Loggers that record trip information for download. Such GPS tracking is useful for trailblazing, mapping by hikers and cyclists, and the production of geocoded photographs.
- Converged devices, including GPS Phones and GPS cameras, in which GNSS is a feature rather than the main purpose of the device. The majority of GNSS devices is now converged devices, and may use assisted GPS or standalone (not network dependent) or both. The vulnerability of consumer GNSS to radio frequency interference from planned wireless data services is controversial.

Dedicated GNSS Navigation Devices

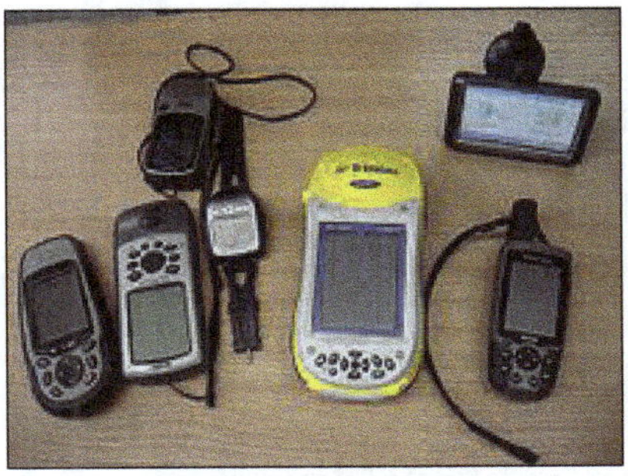

Hand-held receivers.

A taxi equipped with GPS.

Dedicated devices have various degrees of mobility. Hand-held, outdoor, or sport receivers have replaceable batteries that can run them for several hours, making them suitable for hiking, bicycle touring and other activities far from an electric power source. Their screens are small, and some do not show color, in part to save power. Some use transflective liquid-crystal displays, allowing use in bright sunlight. Cases are rugged and some are water-resistant.

Other receivers, often called *mobile* are intended primarily for use in a car, but have a small rechargeable internal battery that can power them for an hour or two away from the car. Special purpose devices for use in a car may be permanently installed and depend entirely on the automotive electrical system.

The pre-installed embedded software of early receivers did not display maps; 21st-century ones commonly show interactive street maps (of certain regions) that may also show points of interest, route information and step-by-step routing directions, often in spoken form with a feature called "text to speech".

Integration with Smartphones

Almost all smartphones now incorporate GNSS receivers. This has been driven both by consumer demand and by service suppliers. There are now many phone apps that depend on location services, such as navigational aids, and multiple commercial opportunities, such as localised advertising. In its early development, access to user location services was driven by European and American emergency services to help locate callers.

All smartphone operating systems offer free mapping and navigational services that require a data connection; some allow the pre-purchase and downloading of maps but the demand for this is diminishing as data connection reliant maps can generally be cached anyway. There are many navigation applications and new versions are constantly being introduced. Major apps include Google Maps Navigation, Apple Maps and Waze, which require data connections, iGo for Android, Maverick and HERE for Windows Phone, which use cached maps and can operate without a data connection. Consequently, almost any smartphone now qualifies as a personal navigation assistant.

The use of mobile phones as navigational devices has outstripped the use of standalone GPS devices. In 2009, independent analyst firm Berg Insight found that GPS-enabled GSM/WCDMA handsets in the USA alone numbered 150 million units, against the sale of only 40 million standalone GPS receivers.

Assisted GPS (A-GPS) uses a combination of satellite data and cell tower data to shorten the time to first fix, reduce the need to download a satellite almanac periodically and to help resolve a location when satellite signals are disturbed by the proximity of large buildings. When out of range of a cell tower the location performance of a phone using A-GPS may be reduced. Phones with an A-GPS based hybrid positioning system can maintain a location fix when GPS signals are inadequate by cell tower triangulation and WiFi hotspot locations. Most smartphones download a satellite almanac when online to accelerate a GPS fix when out of cell tower range.

Some, older, Java-enabled phones lacking integrated GPS may still use external GPS receivers via serial or Bluetooth) connections, but the need for this is now rare.

By tethering to a laptop, some phones can provide localisation services to a laptop as well.

Palm, Pocket and Laptop PC

Software companies have made available GPS navigation software programs for in-vehicle use on laptop computers. Benefits of GPS on a laptop include larger map overview, ability to use the keyboard to control GPS functions, and some GPS software for laptops offers advanced trip-planning features not available on other platforms, such as midway stops, capability of finding alternative scenic routes as well as only highway option.

Palms and Pocket PC's can also be equipped with GPS navigation. A pocket PC differs from a dedicated navigation device as it has an own operating system and can also run other applications.

GPS Modules

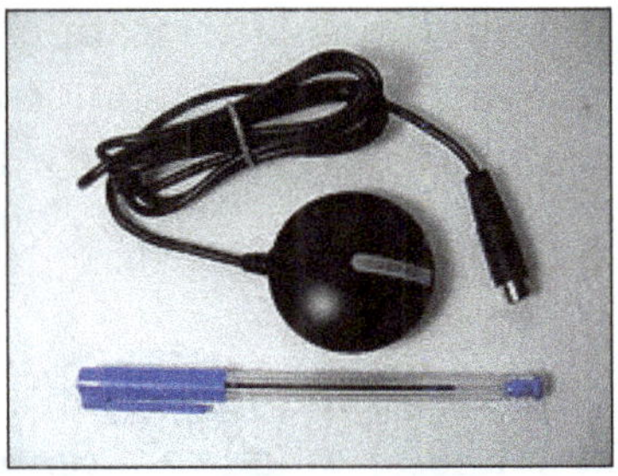

A modern SiRFstarIII chip based 20-channel GPS receiver with WAAS/EGNOS support.

Other GPS devices need to be connected to a computer in order to work. This computer can be a home computer, laptop, PDA, digital camera, or smartphones. Depending on the type of computer and available connectors, connections can be made through a serial or USB cable, as well as Bluetooth, CompactFlash, SD, PCMCIA and the newer ExpressCard. Some PCMCIA/ExpressCard GPS units also include a wireless modem.

Devices usually do not come with pre-installed GPS navigation software, thus, once purchased; the user must install or write their own software. As the user can choose which software to use, it can be better matched to their personal taste. It is very common for a PC-based GPS receiver to come bundled with a navigation software suite. Also, GPS modules are significantly cheaper than complete stand-alone systems. The software may include maps only for a particular region, or the entire world, if software such as Google Maps is used.

Some hobbyists have also made some GPS devices and open-sourced the plans. Examples include the Elektor GPS units. These are based around a SiRFstarIII chip and are comparable to their commercial counterparts. Other chips and software implementations are also available.

Aviators

Aviators use GPS to increase their ability to keep safety up to par and also to maintain the efficiency of the flight. A GPS navigation system can help aviators always know their position and its surroundings in all of its phases starting from its departure all the way to its landing point. Also, a GPS now allows an aviator from start to finish not to have to depend on ground infrastructures and allows them to be able to have a preferred route from its departure and landing point, but not only do they play a part in preferred routes they also help in airports that lack ground-based navigation and surveillance equipment. With the use of a GPS for aviators, it saves time and money being used on fuel. More efficient air routes are continuing to expand every day. There are now some GPS units that allow aviators to get a clearer look in areas where the satellite is augmented to be able to have safe landings in bad visibility conditions. There have now been two new signals made for GPS, the first being made to help in critical conditions in the sky and the other will make GPS more of a robust navigation service. Many aviator services have now made it a required service to use a GPS. Commercial aviation applications include GPS devices that calculate location and feed that information to large multi-input navigational computers for autopilot, course information and correction displays to the pilots, and course tracking and recording devices.

Military

Military applications include devices similar to consumer sport products for foot soldiers (commanders and regular soldiers), small vehicles and ships, and devices similar to commercial aviation applications for aircraft and missiles. Examples are the United States military's Commander's Digital Assistant and the Soldier Digital Assistant. Prior to May 2000 only the military had access to the full accuracy of GPS. Consumer devices were restricted by selective availability (SA), which was scheduled to be phased out but was removed abruptly by President Clinton. Differential GPS is a method of cancelling out the error of SA and improving GPS accuracy, and has been routinely available in commercial applications such as for golf carts. GPS is limited to about 15 meter accuracy even without SA. DGPS can be within a few centimeters.

Sequential Receivers

A sequential GPS receiver tracks the necessary satellites by typically using one or two hardware channels. The set will track one satellite at a time, time tag the measurements and combine them when all four satellite pseudoranges have been measured. These receivers are among the least expensive available, but they cannot operate under high dynamics and have the slowest time-to-first-fix (TTFF) performance.

Mishaps

GPS maps and directions are occasionally imprecise. Some people have gotten lost by asking for the shortest route, like a couple in the United States who were looking for the shortest route from South Oregon to Jackpot, Nevada.

In August 2009 a young mother and her six-year-old son became stranded in Death Valley after following GPS directions that led her up an unpaved dead-end road. When they were found five days later, her son had died from the effects of heat and dehydration.

In May 2012, Japanese tourists in Australia were stranded when traveling to North Stradbroke Island and their GPS receiver instructed them to drive into Moreton Bay.

In 2008 a GPS sent a softball team bus into a 9 ft. tunnel slicing off the top of the bus and the whole team was hospitalized.

A man name Brad Preston from Oregon City, Oregon has been experiencing troubles with GPS for a while. He says five to eight times a week people pull into his driveway because on a GPS it shows a street through his property.

John and Starry Rhodes a couple from Reno, Nevada were driving home from Oregon when they started to see there was a lot of snow in the area but thought to keep going because they were already on the road for 30 miles. But really the GPS led them to a road that was not plowed in the Oregon forest and they were stuck for 3 days.

A lady name Mary Davis was driving in an unfamiliar place when her GPS told her to make a right turn onto a train track while there was a train coming down. Mary was lucky there was a local police officer who noticed the situation and urged her quickly to get out of the car as fast as she could. Mary was lucky enough to get out of the car leaving it for the train to hit it and totaling it. The officer commented after and said there was a very good chance that they could have had a fatality on their hands.

Other hazards involve an alley being listed as a street, a lane being identified as a road, or rail tracks as a road.

Obsolete maps sometimes cause the unit to lead a user on an indirect, time-wasting route, because roads may change over time. Smartphone GPS information is usually updated automatically, and free of additional charge. Manufacturers of separate GPS devices also offer map update services for their merchandise, usually for a fee.

Privacy Concerns

User privacy may be compromised if GPS receivers in handheld devices such as mobile phones upload user geo-location data through associated software installed on the device. User geo-location is currently the basis for navigational apps such as Google Maps, location-based advertising, which can promote nearby shops and may allow an advertising agency to track user movements and habits for future use. Regulatory bodies differ between countries regarding the treatment of geo-location data as privileged or not. Privileged data cannot be stored, or otherwise used, without the user's consent.

GPS vehicle tracking systems allow employers to track their employees' location raising questions regarding violation of employee privacy. There are cases where employers continued to collect geo-location data when an employee was off duty in private time.

Rental car services may use the same technique to geo-fence their customers to the areas they have paid for, charging additional fees for violations. In 2010, New York

Civil Liberties Union filed a case against the Labor Department for firing Michael Cunningham after tracking his daily activity and locations using a GPS device attached to his car. Private investigators use planted GPS devices to provide information to their clients on a target's movements.

GPS Signals

Global Positioning System (GPS) satellites broadcast microwave signals to enable GPS receivers on or near the Earth's surface to determine location and time, and to derive velocity. The system is operated by the U.S. Department of Defense (DoD) for use by both the military and the general public.

GPS signals include ranging signals, used to measure the distance to the satellite, and navigation messages. The navigation messages include *ephemeris* data, used to calculate the position of each satellite in orbit, and information about the time and status of the entire satellite constellation, called the *almanac.*

There are four signals available for civilian use. In order of date of introduction, these are: L1 C/A, L2C, L5 and L1C. L1 C/A is also called the legacy signal and is broadcast by all satellites. The other signals are called modernized signals and are not broadcast by all satellites. In addition, there are restricted signals with published frequencies and chip rates but encrypted coding intended to be used only by authorized parties. Some limited use of restricted signals can still be made by civilians without decryption; this is called codeless and semi-codeless access, and is officially supported.

The interface to the User Segment (GPS receivers) is described in the Interface Control Documents (ICD). The format of civilian signals is described in the Interface Specification (IS) which is a subset of the ICD.

Artist's conception of GPS Block II-F satellite in Earth orbit.

Civilian GPS receiver ("GPS navigation device") in a marine application.

Common Characteristics

The GPS satellites (called space vehicles in the GPS interface specification documents) transmit simultaneously several ranging codes and navigation data using binary phase-shift keying (BPSK). Only a limited number of central frequencies are used; satellites using the same frequency are distinguished by using different ranging codes; in other words, GPS uses code division multiple access. The ranging codes are also called chipping codes (in reference to CDMA/DSSS), pseudorandom noise and pseudorandom binary sequences (in reference to the fact that it is predictable, but statistically it resembles noise).

Some satellites transmit several BPSK streams at the same frequency in quadrature, in a form of quadrature amplitude modulation. However, unlike typical QAM systems where a single bit stream is split in two half-symbol-rate bit streams to improve spectral efficiency, in GPS signals the in-phase and quadrature components are modulated by separate (but functionally related) bit streams.

Satellites are uniquely identified by a serial number called *space vehicle number* (SVN) which does not change during its lifetime. In addition, all operating satellites are numbered with a *space vehicle identifier* (SV ID) and *pseudorandom noise number* (PRN number) which uniquely identifies the ranging codes that a satellite uses. There is a fixed one-to-one correspondence between SV identifiers and PRN numbers described in the interface specification. Unlike SVNs, the SV ID/PRN number of a satellite may be changed (also changing the ranging codes it uses). At any point in time, any SV ID/PRN number is in use by at most a single satellite. A single SV ID/PRN number may have been used by several satellites at different points in time and a single satellite may have used different SV ID/PRN numbers at different points in time. The current SVNs and PRN numbers for the GPS constellation may be found at NAVCEN.

The original GPS design contains two ranging codes: the *coarse/acquisition* (C/A) code, which is freely available to the public, and the restricted *precision* (P) code, usually reserved for military applications.

Coarse/Acquisition Code

The C/A PRN codes are Gold codes with a period of 1023 chips transmitted at 1.023 Mchip/s, causing the code to repeat every 1 millisecond. They are exclusive-ored with a 50 bit/s navigation message and the result phase modulates the carrier as previously described. These codes only match up, or strongly autocorrelate when they are almost exactly aligned. Each satellite uses a unique PRN code, which does not correlate well with any other satellite's PRN code. In other words, the PRN codes are highly orthogonal to one another. The 1 ms period of the C/A code corresponds to 299.8 km of distance, and each chip corresponds to a distance of 293 m. (Receivers track these codes well within one chip of accuracy, so measurement errors are considerably smaller than 293 m.)

The C/A codes are generated by combining (using "exclusive or") 2-bit streams generated by maximal period 10 stage linear feedback shift registers (LFSR). Different codes are obtained by selectively delaying one of those bit streams. Thus:

$$C/A_i(t) = A(t) \oplus B(t-D_i)$$

where:

C/A_i is the code with PRN number i.

A is the output of the first LFSR whose generator polynomial is $x \to x^{10} + x^3 + 1$, and initial state is 1111111111_2.

B is the output of the second LFSR whose generator polynomial is $x \to x^{10} + x^9 + x^8 + x^6 + x^3 + x^2 + 1$ and initial state is also 1111111111_2.

D_i is a delay (by an integer number of periods) specific to each PRN number i; it is designated in the GPS interface specification.

$\oplus$ is exclusive or.

The arguments of the functions therein are the number of *bits* or *chips* since their epochs, starting at 0. The epoch of the LFSRs is the point at which they are at the initial state; and for the overall C/A codes it is the start of any UTC second plus any integer number of milliseconds. The output of LFSRs at negative arguments is defined consistent with the period which is 1,023 chips (this provision is necessary because B may have a negative argument using the above equation).

The delay for PRN numbers 34 and 37 is the same; therefore their C/A codes are identical and are not transmitted at the same time (it may make one or both of those signals unusable due to mutual interference depending on the relative power levels received on each GPS receiver).

Precision Code

The P-code is a PRN sequence much longer than the C/A code: $6.187104 \cdot 10^{12}$ chips

(773,388 MByte). Even though the P-code chip rate (10.23 Mchips/s) is ten times that of the C/A code, it repeats only once per week, eliminating range ambiguity. It was assumed that receivers could not directly acquire such a long and fast code so they would first "bootstrap" themselves with the C/A code to acquire the spacecraft ephemerides (positions), produce an approximate time and position fix, and then acquire the P-code to refine the fix.

Whereas the C/A PRNs are unique for each satellite, each satellite transmits a different segment of a master P-code sequence approximately $2.35 \cdot 10^{14}$ chips long (235,000,000,000,000 bits, ~26.716 terabytes). Each satellite repeatedly transmits its assigned segment of the master code, restarting every Sunday at 00:00:00 GPS time.

The P code is public, so to prevent unauthorized users from using or potentially interfering with it through spoofing, the P-code is XORed with *W-code*, a cryptographically generated sequence, to produce the *Y-code*. The Y-code is what the satellites have been transmitting since the anti-spoofing module was set to the "on" state. The encrypted signal is referred to as the *P(Y)-code*.

The details of the W-code are secret, but it is known that it is applied to the P-code at approximately 500 kHz, about 20 times slower than the P-code chip rate. This has led to semi-codeless approaches for tracking the P(Y) signal without knowing the W-code.

Navigation Message

GPS message format		
Sub-frame	Word	Description
1	1–2	Telemetry and handover words (TLM and HOW)
	3–10	Satellite clock, GPS time relationship
2–3	1–2	Telemetry and handover words (TLM and HOW)
	3–10	Ephemeris (precise satellite orbit)
4–5	1–2	Telemetry and handover words (TLM and HOW)
	3–10	Almanac component (satellite network synopsis, error correction)

In addition to the PRN ranging codes, a receiver needs to know the time and position of each active satellite. GPS encodes this information into the *navigation message* and modulates it onto both the C/A and P(Y) ranging codes at 50 bit/s. The navigation message format is called LNAV data (for *legacy navigation*).

The navigation message conveys information of three types:

- The GPS date and time and the satellite's status.

- The ephemeris: Precise orbital information for the transmitting satellite.
- The almanac: Status and low-resolution orbital information for every satellite.

An ephemeris is valid for only four hours; an almanac is valid with little dilution of precision for up to two weeks. The receiver uses the almanac to acquire a set of satellites based on stored time and location. As each satellite is acquired, its ephemeris is decoded so the satellite can be used for navigation.

The navigation message consists of 30-second *frames* 1,500 bits long, divided into five 6-second *subframes* of ten 30-bit words each. Each subframe has the GPS time in 6-second increments. Subframe 1 contains the GPS date (week number) and satellite clock correction information, satellite status and health. Subframes 2 and 3 together contain the transmitting satellite's ephemeris data. Subframes 4 and 5 contain *page* 1 through 25 of the 25-page almanac. The almanac is 15,000 bits long and takes 12.5 minutes to transmit.

A frame begins at the start of the GPS week and every 30 seconds thereafter. Each week begins with the transmission of almanac page 1.

There are two navigation message types: LNAV-L is used by satellites with PRN numbers 1 to 32 (called *lower PRN numbers*) and LNAV-U is used by satellites with PRN numbers 33 to 63 (called *upper PRN numbers*). The 2 types use very similar formats. Subframes 1 to 3 are the same while subframes 4 and 5 are almost the same. Each message type contains almanac data for all satellites using the same navigation message type, but not the other.

Each subframe begins with a Telemetry Word (TLM) that enables the receiver to detect the beginning of a subframe and determine the receiver clock time at which the navigation subframe begins. Next is the handover word (HOW) giving the GPS time (actually the time when the first bit of the next subframe will be transmitted) and identifies the specific subframe within a complete frame. The remaining eight words of the subframe contain the actual data specific to that subframe. Each word includes 6 bits of parity generated using an algorithm based on Hamming codes, which take into account the 24 non-parity bits of that word and the last 2 bits of the previous word.

After a subframe has been read and interpreted, the time the next subframe was sent can be calculated through the use of the clock correction data and the HOW. The receiver knows the receiver clock time of when the beginning of the next subframe was received from detection of the Telemetry Word thereby enabling computation of the transit time and thus the pseudorange.

Time

GPS time is expressed with a resolution of 1.5 seconds as a week number and a time of week count (TOW). Its zero point (week 0, TOW 0) is defined to be 1980-01-06T00:00Z.

The TOW count is a value ranging from 0 to 403,199 whose meaning is the number of 1.5 second periods elapsed since the beginning of the GPS week. Expressing TOW count thus requires 19 bits (2^{19} = 524,288). GPS time is a continuous time scale in that it does not include leap seconds; therefore the start/end of GPS weeks may differ from that of the corresponding UTC day by an integer number of seconds.

In each subframe, each hand-over word (HOW) contains the most significant 17 bits of the TOW count corresponding to the start of the next following subframe. Note that the 2 least significant bits can be safely omitted because one HOW occurs in the navigation message every 6 seconds, which is equal to the resolution of the truncated TOW count thereof. Equivalently, the truncated TOW count is the time duration since the last GPS week start/end to the beginning of the next frame in units of 6 seconds.

Each frame contains (in subframe 1) the 10 least significant bits of the corresponding GPS week number. Note that each frame is entirely within one GPS week because GPS frames do not cross GPS week boundaries. Since rollover occurs every 1,024 GPS weeks (approximately every 19.6 years; 1,024 is 2^{10}), a receiver that computes current calendar dates needs to deduce the upper week number bits or obtain them from a different source. One possible method is for the receiver to save its current date in memory when shut down, and when powered on, assume that the newly decoded truncated week number corresponds to the period of 1,024 weeks that starts at the last saved date. This method correctly deduces the full week number if the receiver is never allowed to remain shut down (or without a time and position fix) for more than 1,024 weeks (~19.6 years).

Almanac

The *almanac* consists of coarse orbit and status information for each satellite in the constellation, an ionospheric model, and information to relate GPS derived time to Coordinated Universal Time (UTC). Each frame contains a part of the almanac (in subframes 4 and 5) and the complete almanac is transmitted by each satellite in 25 frames total (requiring 12.5 minutes). The almanac serves several purposes. The first is to assist in the acquisition of satellites at power-up by allowing the receiver to generate a list of visible satellites based on stored position and time, while an ephemeris from each satellite is needed to compute position fixes using that satellite. In older hardware, lack of an almanac in a new receiver would cause long delays before providing a valid position, because the search for each satellite was a slow process. Advances in hardware have made the acquisition process much faster, so not having an almanac is no longer an issue. The second purpose is for relating time derived from the GPS (called GPS time) to the international time standard of UTC. Finally, the almanac allows a single-frequency receiver to correct for ionospheric delay error by using a global ionospheric model. The corrections are not as accurate as GNSS augmentation systems like WAAS or dual-frequency receivers. However, it is often better than no correction, since ionospheric error is the largest error source for a single-frequency GPS receiver.

Structure of Subframes 4 and 5

LNAV-L frames 4 and 5		
Sub-frame	Page	Description
4	1, 6, 11–12, 16, 19–24	Reserved
	2–5, 7–10	Almanac data for SV 25–32
	13	Navigation message correction table (NMCT)
	14–15	Reserved for system use
	17	Special messages
	18	Ionospheric correction data and UTC
	25	A-S flags for SV 1–32, health info. for SV 25–32
5	1–24	Almanac data for SV 1–24
	25	Health info. for SV 1–24 almanac reference time

LNAV-U frames 4 and 5		
Sub-frame	Page	Description
4	1, 6, 10–12, 16, 19–24	Reserved
	2–5, 7–9	Almanac data for SV 89–95
	13	Navigation message correction table (NMCT)
	14–15	Reserved for system use
	17	Special messages
	18	Ionospheric correction data and UTC
	25	A-S flags for PRN numbers 33–63, health info. for SV 89–95
5	1–24	Almanac data for SV 65–88
	25	Health info. for SV 65–88 almanac reference time

Data Updates

Satellite data is updated typically every 24 hours, with up to 60 days data loaded in case there is a disruption in the ability to make updates regularly. Typically the updates contain new ephemerides, with new almanacs uploaded less frequently. The Control Segment guarantees that during normal operations a new almanac will be uploaded at least every 6 days.

Satellites broadcast a new ephemeris every two hours. The ephemeris is generally valid for 4 hours, with provisions for updates every 4 hours or longer in non-nominal conditions. The time needed to acquire the ephemeris is becoming a significant element of the delay to first position fix, because as the receiver hardware becomes more capable, the time to lock onto the satellite signals shrinks; however, the ephemeris data requires 18 to 36 seconds before it is received, due to the low data transmission rate.

Frequency Information

For the ranging codes and navigation message to travel from the satellite to the receiver, they must be modulated onto a carrier wave. In the case of the original GPS design, two frequencies are utilized; one at 1575.42 MHz (10.23 MHz × 154) called L1; and a second at 1227.60 MHz (10.23 MHz × 120), called L2.

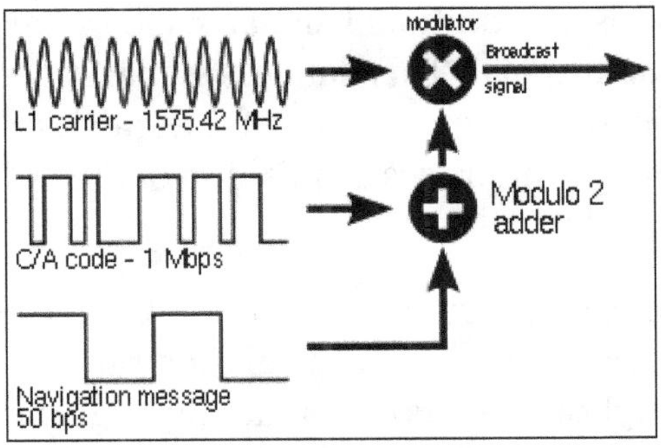

GPS broadcast signal.

The C/A code is transmitted on the L1 frequency as a 1.023 MHz signal using a bi-phase shift keying (BPSK) modulation technique. The P(Y)-code is transmitted on both the L1 and L2 frequencies as a 10.23 MHz signal using the same BPSK modulation, however the P(Y)-code carrier is in quadrature with the C/A carrier (meaning it is 90° out of phase).

Besides redundancy and increased resistance to jamming, a critical benefit of having two frequencies transmitted from one satellite is the ability to measure directly, and therefore remove, the ionospheric delay error for that satellite. Without such a measurement, a GPS receiver must use a generic model or receive ionospheric corrections from another source (such as the Wide Area Augmentation System or WAAS). Advances in the technology used on both the GPS satellites and the GPS receivers has made ionospheric delay the largest remaining source of error in the signal. A receiver capable of performing this measurement can be significantly more accurate and is typically referred to as a *dual frequency receiver*.

Modernization and Additional GPS Signals

Having reached full operational capability on July 17, 1995 the GPS system had completed its original design goals. However, additional advances in technology and new demands on the existing system led to the effort to "modernize" the GPS system. Announcements from the Vice President and the White House in 1998 heralded the beginning of these changes and in 2000, the U.S. Congress reaffirmed the effort, referred to as *GPS III*.

The project involves new ground stations and new satellites, with additional navigation signals for both civilian and military users, and aims to improve the accuracy and availability for all users. A goal of 2013 has been established with incentives offered to the contractors if they can complete it by 2011.

General Features

Modernized GPS civilian signals have two general improvements over their legacy counterparts: a dataless acquisition aid and forward error correction (FEC) coding of the NAV message.

A dataless acquisition aid is an additional signal, called a pilot carrier in some cases, broadcast alongside the data signal. This dataless signal is designed to be easier to acquire than the data encoded and, upon successful acquisition, can be used to acquire the data signal. This technique improves acquisition of the GPS signal and boosts power levels at the correlator.

The second advancement is to use forward error correction (FEC) coding on the NAV message itself. Due to the relatively slow transmission rate of NAV data (usually 50 bits per second), small interruptions can have potentially large impacts. Therefore, FEC on the NAV message is a significant improvement in overall signal robustness.

L2C

One of the first announcements was the addition of a new civilian-use signal, to be transmitted on a frequency other than the L1 frequency used for the coarse/acquisition (C/A) signal. Ultimately, this became the L2C signal, so called because it is broadcast on the L2 frequency. Because it requires new hardware on board the satellite, it is only transmitted by the so-called Block IIR-M and later design satellites. The L2C signal is tasked with improving accuracy of navigation, providing an easy to track signal, and acting as a redundant signal in case of localized interference.

Unlike the C/A code, L2C contains two distinct PRN code sequences to provide ranging information; the *civil-moderate* code (called CM), and the *civil-long* length code (called CL). The CM code is 10,230 bits long, repeating every 20 ms. The CL code is 767,250 bits long, repeating every 1,500 ms. Each signal is transmitted at 511,500 bits per second (bit/s); however, they are multiplexed together to form a 1,023,000-bit/s signal.

CM is modulated with the CNAV Navigation Message, whereas CL does not contain any modulated data and is called a *dataless sequence*. The long, dataless sequence provides for approximately 24 dB greater correlation (~250 times stronger) than L1 C/A-code.

When compared to the C/A signal, L2C has 2.7 dB greater data recovery and 0.7 dB greater carrier-tracking, although its transmission power is 2.3 dB weaker.

CM and CL Codes

The civil-moderate and civil-long ranging codes are generated by a modular LFSR which is reset periodically to a predetermined initial state. The period of the CM and CL is determined by this resetting and not by the natural period of the LFSR (as is the case with the C/A code). The initial states are designated in the interface specification and are different for different PRN numbers and for CM/CL. The feedback polynomial/mask is the same for CM and CL. The ranging codes are thus given by:

$$\mathrm{CM}_i(t) = A(X_i, t \bmod 10\,230)$$

$$\mathrm{CL}_i(t) = A(Y_i, t \bmod 767\,250)$$

where:

CM_i and CL_i are the ranging codes for PRN number i and their arguments are the integer number of chips elapsed (starting at 0) since start/end of GPS week, or equivalently since the origin of the GPS time scale.

$A(x, t)$ is the output of the LFSR when initialized with initial state x after being clocked t times.

X_i and Y_i are the initial states for CM and CL respectively for PRN number i. mod is the remainder of division operation.

t is the integer number of CM and CL chip periods since the origin of GPS time or equivalently, since any GPS second (starting from 0).

The initial states are described in the GPS interface specification as numbers expressed in octal following the convention that the LFSR state is interpreted as the binary representation of a number where the output bit is the least significant bit, and the bit where new bits are shifted in is the most significant bit. Using this convention, the LFSR shifts from most significant bit to least significant bit and when seen in big endian order, it shifts to the right. The states called *final state* in the IS are obtained after 10229 cycles for CM and after 767249 cycles for LM (just before reset in both cases).

The feedback bit mask is $100100101001001010100111100_2$. Again with the convention that the least significant bit is the output bit of the LFSR and the most significant bit is the shift-in bit of the LFSR, 0 means no feedback *into* that position, and 1 means feedback *into* that position.

CNAV Navigation Message

Message Structure (Common Fields)	
Bits	Information
1–8	Preamble
9–14	PRN of transmitting satellite
15–20	Message type ID
21–37	Truncated TOW count
38	Alert flag
277–300	Cyclic redundancy check

Message Types	
Type ID	Description
10–11	Ephemeris and health

12, 31, 37	Almanac parameters
13–14, 34	Differential correction
15, 36	Text messages
30	Ionospheric and group delay correction
32	Earth orientation parameters
33	UTC parameters
35	GPS/GNSS time offset

The CNAV data is an upgraded version of the original NAV navigation message. It contains higher precision representation and nominally more accurate data than the NAV data. The same type of information (time, status, ephemeris, and almanac) is still transmitted using the new CNAV format; however, instead of using frame/sub-frame architecture, it uses a new pseudo-packetized format made of 12-second 300-bit *messages* analogous to LNAV frames. While LNAV frames have a fixed information content, CNAV messages may be of one of several defined types. The type of a frame determines its information content. Messages do not follow a fixed schedule regarding which message types will be used, allowing the Control Segment some versatility. However, for some message types there are lower bounds on how often they will be transmitted.

In CNAV, at least 1 out of every 4 packets is ephemeris data and the same lower bound applies for clock data packets. The design allows for a wide variety of packet types to be transmitted. With a 32-satellite constellation, and the current requirements of what needs to be sent, less than 75% of the bandwidth is used. Only a small fraction of the available packet types have been defined; this enables the system to grow and incorporate advances without breaking compatibility.

There are many important changes in the new CNAV message:

- It uses forward error correction (FEC) provided by a rate 1/2 convolutional code, so while the navigation message is 25-bit/s, a 50-bit/s signal is transmitted.
- Messages carry a 24-bit CRC, against which integrity can be checked.
- The GPS week number is now represented as 13 bits, or 8192 weeks, and only repeats every 157.0 years, meaning the next return to zero won't occur until the year 2137. This is longer compared to the L1 NAV message's use of a 10-bit week number, which returns to zero every 19.6 years.
- There is a packet that contains a GPS-to-GNSS time offset. This allows better interoperability with other global time-transfer systems, such as Galileo and GLONASS, both of which are supported.
- The extra bandwidth enables the inclusion of a packet for differential correction, to be used in a similar manner to satellite based augmentation systems and which can be used to correct the L1 NAV clock data.

- Every packet contains an alert flag, to be set if the satellite data cannot be trusted. This means users will know within 12 seconds if a satellite is no longer usable. Such rapid notification is important for safety-of-life applications, such as aviation.

- Finally, the system is designed to support 63 satellites, compared with 32 in the L1 NAV message.

CNAV messages begin and end at start/end of GPS week plus an integer multiple of 12 seconds. Specifically, the beginning of the first bit (with convolution encoding already applied) to contain information about a message matches the aforesaid synchronization. CNAV messages begin with an 8-bit preamble which is a fixed bit pattern and whose purpose is to enable the receiver to detect the beginning of a message.

Forward Error Correction Code

The convolutional code used to encode CNAV is described by:

$$X_1(t) = d(t) \oplus d(t-2) \oplus d(t-3) \oplus d(t-5) \oplus d(t-6)$$
$$X_2(t) = d(t) \oplus d(t-1) \oplus d(t-2) \oplus d(t-3) \oplus d(t-6)$$

$$d'(t') = \begin{cases} X_1\left(\dfrac{t'}{2}\right) & \text{if } t' \equiv 0 \pmod 2 \\ X_2\left(\dfrac{t'-1}{2}\right) & \text{if } t' \equiv 1 \pmod 2 \end{cases}$$

where:

X_1 and X_2 are the unordered outputs of the convolutional encoder

d is the raw (non FEC encoded) navigation data, consisting of the simple concatenation of the 300-bit messages.

t is the integer number of *non FEC encoded* navigation data bits elapsed since an arbitrary point in time (starting at 0).

d' is the FEC encoded navigation data.

t' is the integer number of *FEC encoded* navigation data bits elapsed since the same epoch than t (likewise starting at 0).

Since the FEC encoded bit stream runs at 2 times the rate than the non FEC encoded bit as already described, then $t = \lfloor \frac{t'}{2} \rfloor$. FEC encoding is performed independently of navigation message boundaries; this follows from the above equations.

L2C Frequency Information

An immediate effect of having two civilian frequencies being transmitted is the civilian receivers can now directly measure the ionospheric error in the same way as dual frequency P(Y)-code receivers. However, users utilizing the L2C signal alone, can expect 65% more position uncertainty due to ionospheric error than with the L1 signal alone.

Military (M-code)

A major component of the modernization process is a new military signal. Called the Military code, or M-code, it was designed to further improve the anti-jamming and secure access of the military GPS signals.

Very little has been published about this new, restricted code. It contains a PRN code of unknown length transmitted at 5.115 MHz. Unlike the P(Y)-code, the M-code is designed to be autonomous, meaning that a user can calculate their position using only the M-code signal. From the P(Y)-code's original design, users had to first lock onto the C/A code and then transfer the lock to the P(Y)-code. Later, direct-acquisition techniques were developed that allowed some users to operate autonomously with the P(Y)-code.

MNAV Navigation Message

A little more is known about the new navigation message, which is called *MNAV*. Similar to the new CNAV, this new MNAV is packeted instead of framed, allowing for very flexible data payloads. Also like CNAV it can utilize Forward Error Correction (FEC) and advanced error detection (such as a CRC).

M-code Frequency Information

The M-code is transmitted in the same L1 and L2 frequencies already in use by the previous military code, the P(Y)-code. The new signal is shaped to place most of its energy at the edges (away from the existing P(Y) and C/A carriers).

In a major departure from previous GPS designs, the M-code is intended to be broadcast from a high-gain directional antenna, in addition to a full-Earth antenna. This directional antenna's signal, called a spot beam, is intended to be aimed at a specific region (several hundred kilometers in diameter) and increase the local signal strength by 20 dB, or approximately 100 times stronger. A side effect of having two antennas is that the GPS satellite will appear to be two GPS satellites occupying the same position to those inside the spot beam. While the whole Earth M-code signal is available on the Block IIR-M satellites, the spot beam antennas will not be deployed until the Block III satellites are deployed, which began in December 2018.

An interesting side effect of having each satellite transmit four separate signals is that the MNAV can potentially transmit four different data channels, offering increased data bandwidth.

The modulation method is binary offset carrier, using a 10.23 MHz subcarrier against the 5.115 MHz code. This signal will have an overall bandwidth of approximately 24 MHz, with significantly separated sideband lobes. The sidebands can be used to improve signal reception.

L5

The L5 signal provides a means of radionavigation secure and robust enough for life critical applications, such as aircraft precision approach guidance. The signal is broadcast in a frequency band protected by the ITU for aeronautical radionavigation services. It was first demonstrated from satellite USA-203 (Block IIR-M), and is available on all satellites from GPS IIF. The L5 band provides additional robustness in the form of interference mitigation, the band being internationally protected, redundancy with existing bands, geostationary satellite augmentation, and ground-based augmentation. The added robustness of this band also benefits terrestrial applications.

Two PRN ranging codes are transmitted on L5 in quadrature: the in-phase code (called *I5-code*) and the quadrature-phase code (called *Q5-code*). Both codes are 10,230 bits long, transmitted at 10.23 MHz (1 ms repetition period), and are generated identically (differing only in initial states). Then, I5 is modulated (by exclusive-or) with navigation data (called L5 CNAV) and a 10-bit Neuman-Hofman code clocked at 1 kHz. Similarly, the Q5-code is then modulated but with only a 20-bit Neuman-Hofman code that is also clocked at 1 kHz.

Compared to L1 C/A and L2, these are some of the changes in L5:

- Improved signal structure for enhanced performance.
- Higher transmitted power than L1/L2 signal (~3 dB, or 2× as powerful).
- Wider bandwidth provides a 10× processing gain, provides sharper autocorrelation (in absolute terms, not relative to chip time duration) and requires a higher sampling rate at the receiver.
- Longer spreading codes (10× longer than C/A).
- Uses the Aeronautical Radionavigation Services band.

I5 and Q5 Codes

The I5-code and Q5-code are generated using the same structure but with different parameters. These codes are the combination (by exclusive-or) of the output of 2 differing linear-feedback shift registers (LFSRs) which are selectively reset:

$$5_i(t) = U(t) \oplus V_i(t)$$

$$U(t) = XA((t \bmod 10\,230) \bmod 8\,190)$$

$$V_i(t) = XB_i(X_i, t \bmod 10\,230)$$

where:

i is an ordered pair (P, n) where $P \in \{I, Q\}$ for in-phase and quadrature-phase, and n a PRN number; both phases and a single PRN are required for the L5 signal from a single satellite.

5_i is the ranging codes for i; also denoted as $I5_n$ and $Q5_n$.

U and V_i are intermediate codes, with U not depending on phase *or* PRN.

The output of two 13-stage LFSRs with clock state t' is used:

$XA(x,t')$ has feedback polynomial $x^{13} + x^{12} + x^{10} + x^9 + 1$, and initial state 1111111111111_2.

$XB_i(x,t')$ has feedback polynomial $x^{13} + x^{12} + x^8 + x^7 + x^6 + x^4 + x^3 + x + 1$, and initial state X_i.

X_i is the initial state specified for the phase and PRN number given by i (designated in the IS).

t is the integer number of chip periods since the origin of GPS time or equivalently, since any GPS second (starting from 0).

A and B are maximal length LFSRs. The modulo operations correspond to resets. Note that both are reset each millisecond (synchronized with C/A code epochs). In addition, the extra modulo operation in the description of A is due to the fact it is reset 1 cycle before its natural period (which is 8,191) so that the next repetition becomes offset by 1 cycle with respect to B (otherwise, since both sequences would repeat, I5 and Q5 would repeat within any 1 ms period as well, degrading correlation characteristics).

L5 Navigation Message

The L5 CNAV data includes SV ephemerides, system time, SV clock behavior data, status messages and time information, etc. The 50 bit/s data is coded in a rate 1/2 convolution coder. The resulting 100 symbols per second (sps) symbol stream is modulo-2 added to the I5-code only; the resultant bit-train is used to modulate the L5 in-phase (I5) carrier. This combined signal is called the L5 Data signal. The L5 quadrature-phase (Q5) carrier has no data and is called the L5 Pilot signal. The format used for L5 CNAV is very similar to that of L2 CNAV. One difference is that it uses 2 times the data rate. The bit fields within each message, message types, and forward error correction code algorithm are the same as those of L2 CNAV. L5 CNAV messages begin and end at start/end of GPS week plus an integer multiple of 6 seconds (this applies to the beginning of the first bit to contain information about a message, as is the case for L2 CNAV).

L5 Frequency Information

Broadcast on the L5 frequency (1176.45 MHz, 10.23 MHz × 115), which is an aeronautical navigation band. The frequency was chosen so that the aviation community can manage interference to L5 more effectively than L2.

L1C

L1C is a civilian-use signal, to be broadcast on the L1 frequency (1575.42 MHz), which contains the C/A signal used by all current GPS users. The L1C will be available with the first Block III launch, tentatively scheduled for the first half of fiscal year 2017.

L1C consists of a pilot (called $L1C_P$) and a data (called $L1C_D$) component. These components use carriers with the same phase (within a margin of error of 100 milliradians), instead of carriers in quadrature as with L5. The PRN codes are 10,230 bits long and transmitted at 1.023 Mbit/s. The pilot component is also modulated by an overlay code called $L1C_O$ (a secondary code that has a lower rate than the ranging code and is also predefined, like the ranging code). Of the total L1C signal power, 25% is allocated to the data and 75% to the pilot. The modulation technique used is BOC(1,1) for the data signal and TMBOC for the pilot. The time multiplexed binary offset carrier (TMBOC) is BOC(1,1) for all except 4 of 33 cycles, when it switches to BOC(6,1).

- Implementation will provide C/A code to ensure backward compatibility.
- Assured of 1.5 dB increase in minimum C/A code power to mitigate any noise floor increase.
- Data-less signal component pilot carrier improves tracking compared with L1 C/A.
- Enables greater civil interoperability with Galileo L1.

L1C Ranging Code

The L1C pilot and data ranging codes are based on a Legendre sequence with length 10223 used to build an intermediate code (called a *Weil code*) which is expanded with a fixed 7-bit sequence to the required 10,230 bits. This 10,230-bit sequence is the ranging code and varies between PRN numbers and between the pilot and data components. The ranging codes are described by:

$$L1C_i(t) = L1C'(t \bmod 10\,230)$$

$$L1C'_i(t') = \begin{cases} W_i(t') & \text{if } t' < p'_i \\ S(t' - p'_i) & \text{if } p'_i \le t' < p'_i + 7 \\ W_i(t' - 7) & \text{if } t' \ge p'_i + 7 \end{cases}$$

$$S = (0,1,1,0,1,0,0)$$

$$W_i(n) = L(n) \oplus L((n + w_i) \bmod 10\,223)$$

$$L(n) = \begin{cases} 1 & \text{if } n \ne 0 \text{ and there is an integer } m \text{ such that } n \equiv m^2 \pmod{10\,223} \\ 0 & \text{otherwise} \end{cases}$$

where:

$L1C_i$ is the ranging code for PRN number and component i.

$L1C'_i$ represents a period of $L1C_i$; it is introduced only to allow a more clear notation. To obtain a direct formula for L1C start from the right side of the formula for $L1C'$ and replace all instances of t' with $t \bmod 10230$.

t is the integer number of L1C chip periods (which is $\frac{1}{1.023}$ µs) since the origin of GPS time or equivalently, since any GPS second (starting from 0).

i is an ordered pair identifying a PRN number and a code ($L1C_P$ or $L1C_D$) and is of the form (P,n) or (D,n) where n is the PRN number of the satellite, and P, D are symbols (not variables) that indicate the $L1C_P$ code or $L1C_D$ code, respectively.

L is an intermediate code: a Legendre sequence whose domain is the set of integers n for which $0 \le n \le 10222$.

W_i is an intermediate code called Weil code, with the same domain as L.

S is a 7-bit long sequence defined for 0-based indexes 0 to 6.

p'_i is the 0-based insertion index of the sequence S into the ranging code (specific for PRN number and code i). It is defined in the Interface Specification (IS) as a 1-based index p, therefore $p'_i = p_i - 1$.

w_i is the Weil index for PRN number and code i designated in the IS.

mod is the remainder of division (or modulo) operation, which differs to the notation in statements of modular congruence.

According to the formula above and the GPS IS, the first w_i bits (equivalently, up to the insertion point of S) of $L1C'_i$ and L1C are the first bits the corresponding Weil code; the next 7 bits are S; the remaining bits are the remaining bits of the Weil code.

The IS asserts that $0 \le p'_i \le 10222$. For clarity, the formula for $L1C'_i$ does not account for the hypothetical case in which $p'_i > 10222$, which would cause the instance of S inserted into $L1C'_i$ to wrap from index 10 229 to 0.

L1C Overlay Code

The overlay codes are 1,800 bits long and are transmitted at 100 bit/s, synchronized with the navigation message encoded in $L1C_D$.

For PRN numbers 1 to 63 they are the truncated outputs of maximal period LFSRs which vary in initial conditions and feedback polynomials.

For PRN numbers 64 to 210 they are truncated Gold codes generated by combining 2

LFSR outputs ($S1_i$ and $S2_i$, where i is the PRN number) whose initial state varies. $S1_i$ has one of the 4 feedback polynomials used overall (among PRN numbers 64–210). $S2_i$ has the same feedback polynomial for all PRN numbers in the range 64–210.

CNAV-2 Navigation Message

Subframes			
Subframe	Bit count		Description
	Raw	Encoded	
1	9	52	Time of interval (TOI)
2	576	1,200	Time correction and ephemeris data
3	250	548	Variable data

Subframe 3 pages	
Page no.	Description
1	UTC & IONO
2	GGTO & EOP
3	Reduced almanac
4	Midi almanac
5	Differential correction
6	Text

The L1C navigation data (called CNAV-2) is broadcast in 1,800 bits long (including FEC) frames and is transmitted at 100 bit/s.

The frames of L1C are analogous to the messages of L2C and L5. While L2 CNAV and L5 CNAV use a dedicated message type for ephemeris data, all CNAV-2 frames include that information.

The common structure of all messages consists of 3 frames, as listed in the adjacent table. The content of subframe 3 varies according to its page number which is analogous to the type number of L2 CNAV and L5 CNAV messages. Pages are broadcast in an arbitrary order.

The time of messages is expressed in a different format than the format of the previous civilian signals. Instead it consists of 3 components:

- The week number, with the same meaning as with the other civilian signals. Each message contains the week number modulo 8,192 or equivalently, the 13 least significant bits of the week number, allowing direct specification of any date within a cycling 157-year range.

- An interval time of week (ITOW): The integer number of 2 hour periods elapsed since the latest start/end of week. It has range 0 to 83 (inclusive), requiring 7 bits to encode.
- A time of interval (TOI): The integer number of 18 second periods elapsed since the period represented by the current ITOW to the beginning of the *next* message. It has range 0 to 399 (inclusive) and requires 9 bits of data.

TOI is the only content of subframe 1. The week number and ITOW are contained in subframe 2 along with other information.

Subframe 1 is encoded by a modified BCH code. Specifically, the 8 least significant bits are BCH encoded to generate 51 bits, then combined using exclusive or with the most significant bit and finally the most significant bit is appended as the most significant bit of the previous result to obtain the final 52 bits. Subframes 2 and 3 are individually expanded with a 24-bit CRC, then individually encoded using a low-density parity-check code, and then interleaved as a single unit using a block interleaver.

Overview of Frequencies

<table>
<tr><th colspan="5">GPS Frequencies</th></tr>
<tr><th>Band</th><th>Frequency (MHz)</th><th>Phase</th><th>Original usage</th><th>Modernized usage</th></tr>
<tr><td rowspan="2">L1</td><td rowspan="2">1575.42
(10.23 × 154)</td><td>I</td><td colspan="2">Encrypted precision P(Y) code</td></tr>
<tr><td>Q</td><td>Coarse/acquisition (C/A) code</td><td>C/A, L1 Civilian (L1C), and Military (M) code</td></tr>
<tr><td rowspan="2">L2</td><td rowspan="2">1227.60
(10.23 × 120)</td><td>I</td><td colspan="2">Encrypted precision P(Y) code</td></tr>
<tr><td>Q</td><td>unmodulated carrier</td><td>L2 Civilian (L2C) code and Military (M) code</td></tr>
<tr><td>L3</td><td>1381.05
(10.23 × 135)</td><td></td><td>used by Nuclear Detonation (NUDET) Detection System Payload (NDS):
signals nuclear detonations/ high-energy infrared events. Used to enforce nuclear test ban treaties.</td><td></td></tr>
<tr><td>L4</td><td>1379.9133
(10.23 × 1214/9)</td><td></td><td>N/A</td><td>being studied for additional ionospheric correction</td></tr>
<tr><td rowspan="2">L5</td><td rowspan="2">1176.45
(10.23 × 115)</td><td>I</td><td rowspan="2">N/A</td><td>Safety-of-Life (SoL) Data signal</td></tr>
<tr><td>Q</td><td>Safety-of-Life (SoL) Pilot signal</td></tr>
</table>

All satellites broadcast at the same two frequencies, 1.57542 GHz (L1 signal) and 1.2276 GHz (L2 signal). The satellite network uses a CDMA spread-spectrum technique where the low-bitrate message data is encoded with a high-rate pseudo-random noise (PRN) sequence that is different for each satellite. The receiver must be aware of the PRN codes for each satellite to reconstruct the actual message data. The C/A code, for civilian use, transmits data at 1.023 million chips per second, whereas the P code, for U.S. military use, transmits at 10.23 million chips per second. The L1 carrier is modulated by both the C/A and P codes, while the L2 carrier is only modulated by the P code. The P code can be encrypted as a so-called P(Y) code which is only available to military equipment with a proper decryption key. Both the C/A and P(Y) codes impart the precise time-of-day to the user.

Each composite signal (in-phase and quadrature phase) becomes:

$$S(t)=\sqrt{P_I}X_I(t)\cos(\omega t+\phi_0)-\sqrt{P_Q}X_Q(t)\underbrace{\sin(\omega t+\phi_0)}_{-\cos\left(\omega t+\phi_0+\frac{\pi}{2}\right)},$$

where P_I and P_Q represent signal powers; $X_I(t)$ and $X_Q(t)$ represent codes with/without data ($=\pm1$). This is a formula for the ideal case (which is not attained in practice) as it does not model timing errors, noise, amplitude mismatch between components or quadrature error (when components are not exactly in quadrature).

Demodulation and Decoding

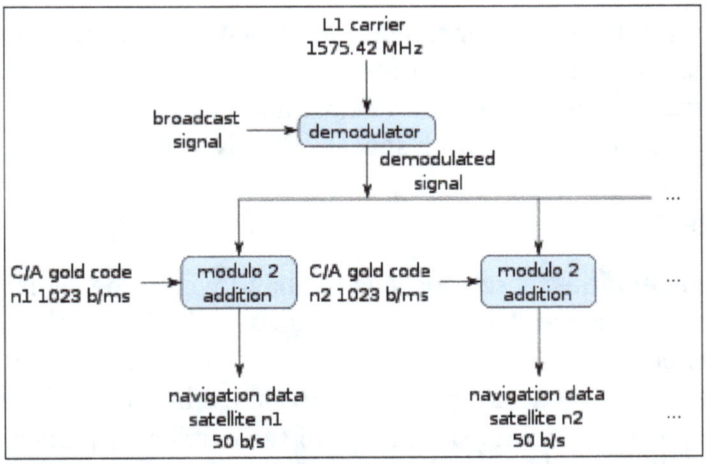

Demodulating and Decoding GPS Satellite Signals using the Coarse/Acquisition Gold code.

A GPS receiver processes the GPS signals received on its antenna to determine position, velocity and/or timing. The signal at antenna is amplified, down converted to baseband or intermediate frequency, filtered (to remove frequencies outside the intended frequency range for the digital signal that would alias into it) and digitalized; these steps may be chained in a different order. Note that aliasing is sometimes intentional (specifically, when undersampling is used) but filtering is still required to discard frequencies not intended to be present in the digital representation.

For each satellite used by the receiver, the receiver must first acquire the signal and then track it as long as that satellite is in use; both are performed in the digital domain in by far most (if not all) receivers.

Acquiring a signal is the process of determining the frequency and code phase (both relative to receiver time) when it was previously unknown. Code phase must be determined within an accuracy that depends on the receiver design (especially the tracking loop); 0.5 times the duration of code chips (approx. 0.489 μs) is a representative value.

Tracking is the process of continuously adjusting the estimated frequency and phase to match the received signal as close as possible and therefore is a phase locked loop. Note that acquisition is performed to start using a particular satellite, but tracking is performed as long as that satellite is in use.

One possible procedure is described for L1 C/A acquisition and tracking, but the process is very similar for the other signals. The described procedure is based on computing the correlation of the received signal with a locally generated replica of the ranging code and detecting the highest peak or lowest valley. The offset of the highest peak or lowest valley contains information about the code phase relative to receiver time. The duration of the local replica is set by receiver design and is typically shorter than the duration of navigation data bits, which is 20 ms.

Acquisition

Acquisition of a given PRN number can be conceptualized as searching for a signal in a bidimensional search space where the dimensions are (1) code phase, (2) frequency. In addition, a receiver may not know which PRN number to search for, and in that case a third dimension is added to the search space: (3) PRN number.

Frequency Space

The frequency range of the search space is the band where the signal may be located given the receiver knowledge. The carrier frequency varies by roughly 5 kHz due to the Doppler effect when the receiver is stationary; if the receiver moves, the variation is higher. The code frequency deviation is 1/1,540 times the carrier frequency deviation for L1 because the code frequency is 1/1,540 of the carrier frequency. The down conversion does not affect the frequency deviation; it only shifts all the signal frequency components down. Since the frequency is referenced to the receiver time, the uncertainty in the receiver oscillator frequency adds to the frequency range of the search space.

Code Phase Space

The ranging code has a period of 1,023 chips each of which lasts roughly 0.977 μs. The code gives strong autocorrelation only at offsets less than 1 in magnitude. The extent of the search space in the code phase dimension depends on the granularity of the offsets at which correlation is computed. It is typical to search for the code phase within a granularity of 0.5 chips or finer; that means 2,046 offsets. There may be more factors increasing the size of the search space of code phase. For example, a receiver may be designed so as to examine 2 consecutive windows of the digitalized signal, so that at least one of them does not contain a navigation bit transition (which worsens the correlation peak); this requires the signal windows to be at most 10 ms long.

PRN Number Space

The lower PRN numbers range from 1 to 32 and therefore there are 32 PRN numbers to search for when the receiver does not have information to narrow the search in this dimension. The higher PRN numbers range from 33 to 66.

If the almanac information has previously been acquired, the receiver picks which satellites to listen for by their PRNs. If the almanac information is not in memory, the receiver enters a search mode and cycles through the PRN numbers until a lock is obtained on one of the satellites. To obtain a lock, it is necessary that there be an unobstructed line of sight from the receiver to the satellite. The receiver can then decode the almanac and determine the satellites it should listen for. As it detects each satellite's signal, it identifies it by its distinct C/A code pattern.

Simple Correlation

The simplest way to acquire the signal (not necessarily the most effective or least computationally expensive) is to compute the dot product of a window of the digitalized signal with a set of locally generated replicas. The locally generated replicas vary in carrier frequency and code phase to cover all the already mentioned search space which is the Cartesian product of the frequency search space and the code phase search space. The carrier is a complex number where real and imaginary components are both sinusoids as described by Euler's formula. The replica that generates the highest magnitude of dot product is likely the one that best matches the code phase and frequency of the signal; therefore, if that magnitude is above a threshold, the receiver proceeds to track the signal or further refine the estimated parameters before tracking. The threshold is used to minimize false positives (apparently detecting a signal when there is in fact no signal), but some may still occur occasionally.

Using a complex carrier allows the replicas to match the digitalized signal regardless of the signal's carrier phase and to detect that phase (the principle is the same used by the Fourier transform). The dot product is a complex number; its magnitude represents the level of similarity between the replica and the signal, as with an ordinary correlation of real-valued time series. The argument of the dot product is an approximation of the corresponding carrier in the digitalized signal.

As an example, assume that the granularity for the search in code phase is 0.5 chips and in frequency is 500 Hz, then there are 1,023/0.5 = 2,046 code phases and 10,000 Hz/500 Hz = 20 frequencies to try for a total of 20×2,046 = 40,920 local replicas. Note that each frequency bin is centered on its interval and therefore covers 250 Hz in each direction; for example, the first bin has a carrier at −4.750 Hz and covers the interval

−5,000 Hz to −4,500 Hz. Code phases are equivalent modulo 1,023 because the ranging code is periodic; for example, phase −0.5 is equivalent to phase 1,022.5.

The following table depicts the local replicas that would be compared against the digitalized signal in this example. "•" means a single local replica while "..". is used for elided local replicas:

Carrier freq. deviation	Code phase (in chips)				
	0.0	0.5	(more phases)	1,022.0	1,022.5
−4,750 Hz	•	•	...	•	•
−4,250 Hz	•	•	...	•	•
(more frequencies)	...	...	...	...	...
4,250 Hz	•	•	...	•	•
4,750 Hz	•	•	...	•	•

Fourier Transform

As an improvement over the simple correlation method, it is possible to implement the computation of dot products more efficiently with a Fourier transform. Instead of performing one dot product for each element in the Cartesian product of code and frequency, a single operation involving FFT and covering all frequencies is performed for each code phase; each such operation is more computationally expensive, but it may still be faster overall than the previous method due to the efficiency of FFT algorithms, and it recovers carrier frequency with a higher accuracy, because the frequency bins are much closely spaced in a DFT.

Specifically, for all code phases in the search space, the digitalized signal window is multiplied element by element with a local replica of the code (with no carrier), then processed with a discrete Fourier transform.

Given the previous example to be processed with this method, assume real-valued data (as opposed to complex data, which would have in-phase and quadrature components), a sampling rate of 5 MHz, a signal window of 10 ms, and an intermediate frequency of 2.5 MHz. There will be 5 MHz×10 ms = 50,000 samples in the digital signal, and therefore 25,001 frequency components ranging from 0 Hz to 2.5 MHz in steps of 100 Hz (note that the 0 Hz component is real because it is the average of a real-valued signal and the 2.5 MHz component is real as well because it is the critical frequency). Only the components (or bins) within 5 kHz of the central frequency are examined, which is the range from 2.495 MHz to 2.505 MHz, and it is covered by 51 frequency components. There are 2,046 code phases as in the previous case, thus in total 51×2,046 = 104,346 complex frequency components will be examined.

Circular Correlation with Fourier Transform

Likewise, as an improvement over the simple correlation method, it is possible to perform a single operation covering all code phases for each frequency bin. The operation performed for each code phase bin involves forward FFT, element-wise multiplication in the frequency domain. inverse FFT, and extra processing so that overall, it computes circular correlation instead of circular convolution. This yields more accurate *code phase determination* than the simple correlation method in contrast with the previous method, which yields more accurate *carrier frequency determination* than the previous method.

Tracking and Navigation Message Decoding

Since the carrier frequency received can vary due to Doppler shift, the points where received PRN sequences begin may not differ from O by an exact integral number of milliseconds. Because of this, carrier frequency tracking along with PRN code tracking are used to determine when the received satellite's PRN code begins. Unlike the earlier computation of offset in which trials of all 1,023 offsets could potentially be required, the tracking to maintain lock usually requires shifting of half a pulse width or less. To perform this tracking, the receiver observes two quantities, phase error and received frequency offset. The correlation of the received PRN code with respect to the receiver generated PRN code is computed to determine if the bits of the two signals are misaligned. Comparisons of the received PRN code with receiver generated PRN code shifted half a pulse width early and half a pulse width late are used to estimate adjustment required. The amount of adjustment required for maximum correlation is used in estimating phase error. Received frequency offset from the frequency generated by the receiver provides an estimate of phase rate error. The command for the frequency generator and any further PRN code shifting required are computed as a function of the phase error and the phase rate error in accordance with the control law used. The Doppler velocity is computed as a function of the frequency offset from the carrier nominal frequency. The Doppler velocity is the velocity component along the line of sight of the receiver relative to the satellite.

As the receiver continues to read successive PRN sequences, it will encounter a sudden change in the phase of the 1,023-bit received PRN signal. This indicates the beginning of a data bit of the navigation message. This enables the receiver to begin reading the 20 millisecond bits of the navigation message. The TLM word at the beginning of each subframe of a navigation frame enables the receiver to detect the beginning of a subframe and determine the receiver clock time at which the navigation subframe begins. The HOW word then enables the receiver to determine which specific subframe is being transmitted. There can be a delay of up to 30 seconds before the first estimate of position because of the need to read the ephemeris data before computing the intersections of sphere surfaces.

After a subframe has been read and interpreted, the time the next subframe was sent can be calculated through the use of the clock correction data and the HOW. The receiver knows the receiver clock time of when the beginning of the next subframe was received from detection of the Telemetry Word thereby enabling computation of the transit time and thus the pseudorange. The receiver is potentially capable of getting a new pseudorange measurement at the beginning of each subframe or every 6 seconds.

Then the orbital position data, or ephemeris, from the navigation message is used to calculate precisely where the satellite was at the start of the message. A more sensitive receiver will potentially acquire the ephemeris data more quickly than a less sensitive receiver, especially in a noisy environment.

Error Analysis for the Global Positioning System

The analysis of errors computed using the Global Positioning System is important for understanding how GPS works, and for knowing what magnitude of errors should be expected. The Global Positioning System makes corrections for receiver clock errors and other effects but there are still residual errors which are not corrected. The Global Positioning System (GPS) was created by the United States Department of Defense (DOD) in the 1970s. It has come to be widely used for navigation both by the U.S. military and the general public.

GPS receiver position is computed based on data received from the satellites. Errors depend on geometric dilution of precision and the sources listed in the table below.

Artist's conception of GPS Block II-F satellite in orbit.

User equivalent range errors (UERE) are shown in the table. There is also a numerical error with an estimated value, σ_{num}, of about 1 meter. The standard deviations, σ_R, for the coarse/acquisition (C/A) and precise codes are also shown in the table. These standard deviations are computed by taking the square root of the sum of the squares of the individual components (i.e., RSS for root sum squares). To get the standard deviation of receiver position estimate, these range errors must be multiplied by the appropriate dilution of precision terms and then RSS'ed with the numerical error. Electronics errors are one of several accuracy-degrading effects outlined in the table above. When taken together, autonomous civilian GPS horizontal position fixes are typically accurate to about 15 meters (50 ft). These effects also reduce the more precise P(Y) code's accuracy. However, the advancement of technology means that in the present, civilian GPS fixes under a clear view of the sky are on average accurate to about 5 meters (16 ft) horizontally.

The term user equivalent range error (UERE) refers to the error of a component in the distance from receiver to a satellite. These UERE errors are given as ± errors thereby implying that they are unbiased or zero mean errors. These UERE errors are therefore used in computing standard deviations. The standard deviation of the error in receiver position, σ_{rc}, is computed by multiplying PDOP (Position Dilution Of Precision) by σ_R, the standard deviation of the user equivalent range errors. σ_R is computed by taking the square root of the sum of the squares of the individual component standard deviations.

PDOP is computed as a function of receiver and satellite positions.

σ_R for the C/A code is given by:

$$3\sigma_R = \sqrt{3^2 + 5^2 + 25^2 + 2^2 + 1^2 + 05^2}\ \text{m} = 6.7\,\text{m}$$

The standard deviation of the error in estimated receiver position σ_{rc}, again for the C/A code is given by:

$$\sigma_{rc} = \sqrt{PDOP^2 \times \sigma_R^2 + \sigma_{num}^2} = \sqrt{PDOP^2 \times 2.2^2 + 1^2}\ \text{m}$$

The error diagram on the left shows the inter relationship of indicated receiver position, true receiver position, and the intersection of the four sphere surfaces.

Sources of User Equivalent Range Errors (UERE)	
Source	Effect (m)
Signal arrival C/A	±3
Signal arrival P(Y)	±0.3
Ionospheric effects	±5
Ephemeris errors	±2.5
Satellite clock errors	±2
Multipath distortion	±1
Tropospheric effects	±0.5
$3\sigma_R$ C/A	±6.7
$3\sigma_R$ P(Y)	±6.0

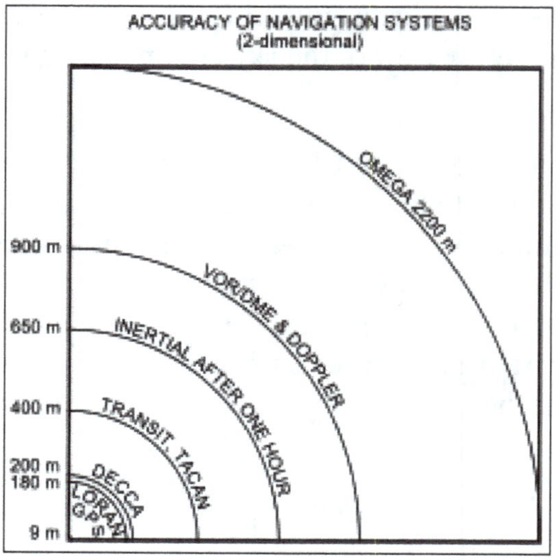

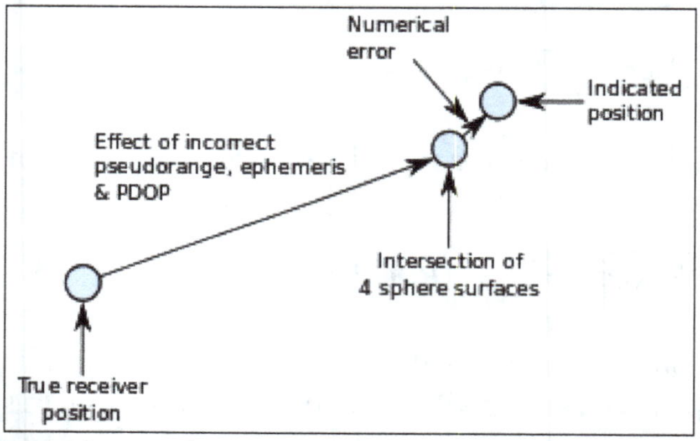

Geometric Error Diagram Showing Typical Relation of Indicated Receiver Position, Intersection of Sphere Surfaces, and True Receiver Position in Terms of Pseudorange Errors, PDOP, and Numerical Errors.

Signal Arrival Time Measurement

The position calculated by a GPS receiver requires the current time, the position of the satellite and the measured delay of the received signal. The position accuracy is primarily dependent on the satellite position and signal delay.

To measure the delay, the receiver compares the bit sequence received from the satellite with an internally generated version. By comparing the rising and trailing edges of the bit transitions, modern electronics can measure signal offset to within about one percent of a bit pulse width, $\frac{0.01\times 300{,}000{,}000\,m/s}{(1.023\times 10^{6}/s)}$, or approximately 10 nanoseconds for the C/A code. Since GPS signals propagate at the speed of light, this represents an error of about 3 meters.

This component of position accuracy can be improved by a factor of 10 using the higher

-chiprate P(Y) signal. Assuming the same one percent of bit pulse width accuracy, the high-frequency P(Y) signal results in an accuracy of $\frac{(0.01\times 300,000,000\ m/s)}{(10.23\times 10^{6}\ /s)}$ or about 30 centimeters.

Atmospheric Effects

Inconsistencies of atmospheric conditions affect the speed of the GPS signals as they pass through the Earth's atmosphere, especially the ionosphere. Correcting these errors is a significant challenge to improving GPS position accuracy. These effects are smallest when the satellite is directly overhead and become greater for satellites nearer the horizon since the path through the atmosphere is longer. Once the receiver's approximate location is known, a mathematical model can be used to estimate and compensate for these errors.

Ionospheric delay of a microwave signal depends on its frequency. It arises from ionized atmosphere. This phenomenon is known as dispersion and can be calculated from measurements of delays for two or more frequency bands, allowing delays at other frequencies to be estimated. Some military and expensive survey-grade civilian receivers calculate atmospheric dispersion from the different delays in the L1 and L2 frequencies, and apply a more precise correction. This can be done in civilian receivers without decrypting the P(Y) signal carried on L2, by tracking the carrier wave instead of the modulated code. To facilitate this on lower cost receivers, a new civilian code signal on L2, called L2C, was added to the Block IIR-M satellites, which was first launched in 2005. It allows a direct comparison of the L1 and L2 signals using the coded signal instead of the carrier wave.

The effects of the ionosphere generally change slowly, and can be averaged over time. Those for any particular geographical area can be easily calculated by comparing the GPS-measured position to a known surveyed location. This correction is also valid for other receivers in the same general location. Several systems send this information over radio or other links to allow L1-only receivers to make ionospheric corrections. The ionospheric data are transmitted via satellite in Satellite Based Augmentation Systems (SBAS) such as Wide Area Augmentation System (WAAS) (available in North America and Hawaii), EGNOS (Europe and Asia), Multi-functional Satellite Augmentation System (MSAS) (Japan), and GPS Aided Geo Augmented Navigation (GAGAN) (India) which transmits it on the GPS frequency using a special pseudo-random noise sequence (PRN), so only one receiver and antenna are required.

Humidity also causes a variable delay, resulting in errors similar to ionospheric delay, but occurring in the troposphere. This effect is more localized than ionospheric effects, changes more quickly and is not frequency dependent. These traits make precise measurement and compensation of humidity errors more difficult than ionospheric effects.

The Atmospheric pressure can also change the signals reception delay, due to the dry gases present at the troposphere (78% N_2, 21% O_2, 0.9% Ar). Its effect varies with local temperature and atmospheric pressure in quite a predictable manner using the laws of the ideal gases.

Multipath Effects

GPS signals can also be affected by multipath issues, where the radio signals reflect off surrounding terrain; buildings, canyon walls, hard ground, etc. These delayed signals cause measurement errors that are different for each type of GPS signal due to its dependency on the wavelength.

A variety of techniques, most notably narrow correlator spacing, have been developed to mitigate multipath errors. For long delay multipath, the receiver itself can recognize the wayward signal and discard it. To address shorter delay multipath from the signal reflecting off the ground, specialized antennas (e.g., a choke ring antenna) may be used to reduce the signal power as received by the antenna. Short delay reflections are harder to filter out because they interfere with the true signal, causing effects almost indistinguishable from routine fluctuations in atmospheric delay.

Multipath effects are much less severe in moving vehicles. When the GPS antenna is moving, the false solutions using reflected signals quickly fail to converge and only the direct signals result in stable solutions.

Ephemeris and Clock Errors

While the ephemeris data is transmitted every 30 seconds, the information itself may be up to two hours old. Variability in solar radiation pressure has an indirect effect on GPS accuracy due to its effect on ephemeris errors. If a fast time to first fix (TTFF) is needed, it is possible to upload a valid ephemeris to a receiver, and in addition to setting the time, a position fix can be obtained in under ten seconds. It is feasible to put such ephemeris data on the web so it can be loaded into mobile GPS devices.

The satellites' atomic clocks experience noise and clock drift errors. The navigation message contains corrections for these errors and estimates of the accuracy of the atomic clock. However, they are based on observations and may not indicate the clock's current state.

These problems tend to be very small, but may add up to a few meters (tens of feet) of inaccuracy.

For very precise positioning (e.g., in geodesy), these effects can be eliminated by differential GPS: the simultaneous use of two or more receivers at several survey points. In the 1990s when receivers were quite expensive, some methods of *quasi-differential* GPS were developed, using only *one* receiver but reoccupation of measuring points. At the TU Vienna the method was named *qGPS* and post processing software was developed.

Geometric Dilution of Precision Computation

Computation of Geometric Dilution of Precision

The concept of geometric dilution of precision was introduced in, *error sources and analysis*. Computations were provided to show how PDOP was used and how it affected the receiver position error standard deviation.

When visible GPS satellites are close together in the sky (i.e., small angular separation), the DOP values are high; when far apart, the DOP values are low. Conceptually, satellites that are close together cannot provide as much information as satellites that are widely separated. Low DOP values represent a better GPS positional accuracy due to the wider angular separation between the satellites used to calculate GPS receiver position. HDOP, VDOP, PDOP and TDOP are respectively Horizontal, Vertical, Position (3-D) and Time Dilution of Precision.

We now take on the task of how to compute the dilution of precision terms. As a first step in computing DOP, consider the unit vector from the receiver to satellite i with components $\frac{(x_i - x)}{R_i}$, $\frac{(y_i - y)}{R_i}$, and $\frac{(z_i - z)}{R_i}$ where the distance from receiver to the satellite, R_i is given by:

$$R_i = \sqrt{(x_i - x)^2 + (y_i - y)^2 + (z_i - z)^2}$$

where x, y, and z denote the position of the receiver and x_i, y_i, and z_i denote the position of satellite i. These x, y, and z components may be components in a North, East, Down coordinate system a South, East, Up coordinate system or other convenient system. Formulate the matrix A as:

$$A = \begin{bmatrix} \frac{(x_1 - x)}{R_1} & \frac{(y_1 - y)}{R_1} & \frac{(z_1 - z)}{R_1} & 1 \\ \frac{(x_2 - x)}{R_2} & \frac{(y_2 - y)}{R_2} & \frac{(z_2 - z)}{R_2} & 1 \\ \frac{(x_3 - x)}{R_3} & \frac{(y_3 - y)}{R_3} & \frac{(z_3 - z)}{R_3} & 1 \\ \frac{(x_4 - x)}{R_4} & \frac{(y_4 - y)}{R_4} & \frac{(z_4 - z)}{R_4} & 1 \end{bmatrix}$$

The first three elements of each row of A are the components of a unit vector from the receiver to the indicated satellite. The elements in the fourth column are c where c denotes the speed of light. Formulate the matrix, Q, as:

$$Q = \left(A^T A\right)^{-1}$$

The weighting matrix, P, has been set to the identity matrix. The elements of the Q matrix are designated as:

$$Q = \begin{bmatrix} d_x^2 & d_{xy}^2 & d_{xz}^2 & d_{xt}^2 \\ d_{xy}^2 & d_y^2 & d_{yz}^2 & d_{yt}^2 \\ d_{xz}^2 & d_{yz}^2 & d_z^2 & d_{zt}^2 \\ d_{xt}^2 & d_{yt}^2 & d_{zt}^2 & d_t^2 \end{bmatrix}$$

The Greek letter σ is used quite often where we have used d. However the elements of the Q matrix do not represent variances and covariances as they are defined in probability and statistics. Instead they are strictly geometric terms. Therefore, d as in dilution of precision is used. PDOP, TDOP and GDOP are given by:

$$PDOP = \sqrt{d_x^2 + d_y^2 + d_z^2}$$

$$TDOP = \sqrt{d_t^2} = |d_t|$$

$$GDOP = \sqrt{PDOP^2 + TDOP^2}$$

The horizontal dilution of precision, $HDOP = \sqrt{d_x^2 + d_y^2}$, and the vertical dilution of precision, $VDOP = \sqrt{d_z^2} = |d_z|$, are both dependent on the coordinate system used. To correspond to the local horizon plane and the local vertical, x, y, and z should denote positions in either a North, East, Down coordinate system or a South, East, Up coordinate system.

Derivation of Equations for Computing Geometric Dilution of Precision

Consider the position error vector, **e** defined as the vector from the intersection of the four sphere surfaces corresponding to the pseudoranges to the true position of the receiver. $\mathbf{e} = e_x\hat{x} + e_y\hat{y} + e_z\hat{z}$ where bold denotes a vector and $\hat{x}$, $\hat{y}$, and $\hat{z}$ denote unit vectors along the x, y, and z axes respectively. Let e_t denote the time error, the true time minus the receiver indicated time. Assume that the mean value of the three components of **e** and e_t are zero.

$$A\begin{bmatrix} e_x \\ e_y \\ e_z \\ e_t \end{bmatrix} = \begin{bmatrix} \frac{(x_1 - x)}{R_1} & \frac{(y_1 - y)}{R_1} & \frac{(z_1 - z)}{R_1} & 1 \\ \frac{(x_2 - x)}{R_2} & \frac{(y_2 - y)}{R_2} & \frac{(z_2 - z)}{R_2} & 1 \\ \frac{(x_3 - x)}{R_3} & \frac{(y_3 - y)}{R_3} & \frac{(z_3 - z)}{R_3} & 1 \\ \frac{(x_4 - x)}{R_4} & \frac{(y_4 - y)}{R_4} & \frac{(z_4 - z)}{R_4} & 1 \end{bmatrix} \begin{bmatrix} e_x \\ e_y \\ e_z \\ e_t \end{bmatrix} = \begin{bmatrix} e_1 \\ e_2 \\ e_3 \\ e_4 \end{bmatrix}$$

where e_1, e_2, e_3, and e_4 are the errors in pseudoranges 1 through 4 respectively. This equation comes from linearizing the Newton-Raphson equation relating pseudoranges to receiver position, satellite positions, and receiver clock errors. Multiplying both sides by A^{-1} there results:

$$\begin{bmatrix} e_x \\ e_y \\ e_z \\ e_t \end{bmatrix} = A^{-1} \begin{bmatrix} e_1 \\ e_2 \\ e_3 \\ e_4 \end{bmatrix}.$$

Transposing both sides:

$$\begin{bmatrix} e_x & e_y & e_z & e_t \end{bmatrix} = \begin{bmatrix} e_1 & e_2 & e_3 & e_4 \end{bmatrix} \left(A^{-1}\right)^t.$$

Post multiplying the matrices on both sides of $\begin{bmatrix} e_x \\ e_y \\ e_z \\ e_t \end{bmatrix} = A^{-1} \begin{bmatrix} e_1 \\ e_2 \\ e_3 \\ e_4 \end{bmatrix}$ by the corresponding matrices in the above equation, there results:

$$\begin{bmatrix} e_x \\ e_y \\ e_z \\ e_t \end{bmatrix} \begin{bmatrix} e_x & e_y & e_z & e_t \end{bmatrix} = A^{-1} \begin{bmatrix} e_1 \\ e_2 \\ e_3 \\ e_4 \end{bmatrix} \begin{bmatrix} e_1 & e_2 & e_3 & e_4 \end{bmatrix} \left(A^{-1}\right)^T.$$

Taking the expected value of both sides and taking the non-random matrices outside the expectation operator, E, there results:

$$E\left(\begin{bmatrix} e_x \\ e_y \\ e_z \\ e_t \end{bmatrix} \begin{bmatrix} e_x & e_y & e_z & e_t \end{bmatrix} \right) = A^{-1} E\left(\begin{bmatrix} e_1 \\ e_2 \\ e_3 \\ e_4 \end{bmatrix} \begin{bmatrix} e_1 & e_2 & e_3 & e_4 \end{bmatrix} \right) \left(A^{-1}\right)^T$$

Assuming the pseudorange errors are uncorrelated and have the same variance, the covariance matrix on the right side can be expressed as a scalar times the identity matrix. Thus:

$$\begin{bmatrix} \sigma_x^2 & \sigma_{xy}^2 & \sigma_{xz}^2 & \sigma_{xt}^2 \\ \sigma_{xy}^2 & \sigma_y^2 & \sigma_{yz}^2 & \sigma_{yt}^2 \\ \sigma_{xz}^2 & \sigma_{yz}^2 & \sigma_z^2 & \sigma_{zt}^2 \\ \sigma_{xt}^2 & \sigma_{yt}^2 & \sigma_{zt}^2 & \sigma_t^2 \end{bmatrix} = \sigma_R^2 \, A^{-1} \left(A^{-1}\right)^T = \sigma_R^2 \left(A^T A\right)^{-1}$$

since $A^{-1}\left(A^{-1}\right)^{T}\left(A^{T}A\right)=I$

Note: $\left(A^{-1}\right)^{T}=\left(A^{T}\right)^{-1}$, since $I=\left(AA^{-1}\right)^{T}=\left(A^{-1}\right)^{T}A^{T}$

Substituting for $\left(A^{T}A\right)^{-1}=Q$ there follows:

$$\begin{bmatrix} \sigma_x^2 & \sigma_{xy}^2 & \sigma_{xz}^2 & \sigma_{xt}^2 \\ \sigma_{xy}^2 & \sigma_y^2 & \sigma_{yz}^2 & \sigma_{yt}^2 \\ \sigma_{xz}^2 & \sigma_{yz}^2 & \sigma_z^2 & \sigma_{zt}^2 \\ \sigma_{xt}^2 & \sigma_{yt}^2 & \sigma_{zt}^2 & \sigma_t^2 \end{bmatrix} = \sigma_R^2 \begin{bmatrix} d_x^2 & d_{xy}^2 & d_{xz}^2 & d_{xt}^2 \\ d_{xy}^2 & d_y^2 & d_{yz}^2 & d_{yt}^2 \\ d_{xz}^2 & d_{yz}^2 & d_z^2 & d_{zt}^2 \\ d_{xt}^2 & d_{yt}^2 & d_{zt}^2 & d_t^2 \end{bmatrix}$$

From the above equation, it follows that the variances of indicated receiver position and time are:

$$\sigma_{rc}^2=\sigma_x^2+\sigma_y^2+\sigma_z^2=\sigma_R^2\left(d_x^2+d_y^2+d_z^2\right)=PDOP^2\sigma_R^2$$

and,

$$\sigma_t^2=\sigma_R^2 d_t^2=TDOP^2\sigma_R^2$$

The remaining position and time error variance terms follow in a straightforward manner.

Selective Availability

GPS included a (currently disabled) feature called *Selective Availability* (*SA*) that adds intentional, time varying errors of up to 100 meters (328 ft.) to the publicly available navigation signals. This was intended to deny an enemy the use of civilian GPS receivers for precision weapon guidance.

SA errors are actually pseudorandom, generated by a cryptographic algorithm from a classified *seed* key available only to authorized users (the U.S. military, its allies and a few other users, mostly government) with a special military GPS receiver. Mere possession of the receiver is insufficient; it still needs the tightly controlled daily key.

Before it was turned off on May 2, 2000, typical SA errors were about 50 m (164 ft.) horizontally and about 100 m (328 ft) vertically. Because SA affects every GPS receiver in a given area almost equally, a fixed station with an accurately known position can measure the SA error values and transmit them to the local GPS receivers so they may correct their position fixes. This is called Differential GPS or DGPS. DGPS also corrects for several other important sources of GPS errors, particularly ionospheric delay, so it continues to be widely used even though SA has been turned off. The ineffectiveness of SA in the face of widely available DGPS was a common argument for turning off SA, and this was finally done by order of President Clinton in 2000.

DGPS services are widely available from both commercial and government sources. The latter include WAAS and the U.S. Coast Guard's network of LF marine navigation beacons. The accuracy of the corrections depends on the distance between the user and the DGPS receiver. As the distance increases, the errors at the two sites will not correlate as well, resulting in less precise differential corrections.

During the 1990–91 Gulf War, the shortage of military GPS units caused many troops and their families to buy readily available civilian units. Selective Availability significantly impeded the U.S. military's own battlefield use of these GPS, so the military made the decision to turn it off for the duration of the war.

In the 1990s, the FAA started pressuring the military to turn off SA permanently. This would save the FAA millions of dollars every year in maintenance of their own radio navigation systems. The amount of error added was "set to zero" at midnight on May 1, 2000 following an announcement by U.S. President Bill Clinton, allowing users access to the error-free L1 signal. Per the directive, the induced error of SA was changed to add no error to the public signals (C/A code). Clinton's executive order required SA to be set to zero by 2006; it happened in 2000 once the U.S. military developed a new system that provides the ability to deny GPS (and other navigation services) to hostile forces in a specific area of crisis without affecting the rest of the world or its own military systems.

On 19 September 2007, the United States Department of Defense announced that future GPS III satellites will not be capable of implementing SA, eventually making the policy permanent.

Anti-spoofing

Another restriction on GPS, antispoofing, remains on. This encrypts the *P-code* so that it cannot be mimicked by a transmitter sending false information. Few civilian receivers have ever used the P-code, and the accuracy attainable with the public C/A code was much better than originally expected (especially with DGPS). So much so that the antispoof policy has relatively little effect on most civilian users. Turning off antispoof would primarily benefit surveyors and some scientists who need extremely precise positions for experiments such as tracking tectonic plate motion.

Relativity

A number of sources of error exist due to relativistic effects that would render the system useless if uncorrected. Three relativistic effects are the time dilation, gravitational frequency shift, and eccentricity effects. Examples include the relativistic time *slowing* due to the speed of the satellite of about 1 part in 10^{10}, the gravitational time dilation that makes a satellite run about 5 parts in 10^{10} *faster* than an Earth-based clock, and the Sagnac effect due to rotation relative to receivers on Earth. These topics are examined below, one at a time.

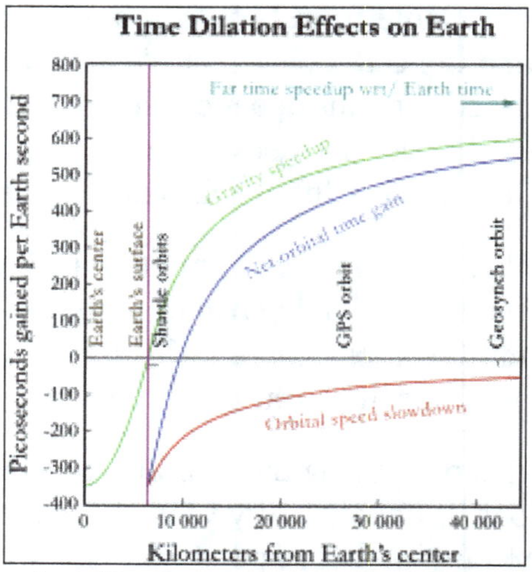

Satellite clocks are slowed by their orbital speed but sped up by their distance out of the Earth's gravitational well.

Special and General Relativity

According to the theory of relativity, due to their constant movement and height relative to the Earth-centered, non-rotating approximately inertial reference frame, the clocks on the satellites are affected by their speed. Special relativity predicts that the frequency of the atomic clocks moving at GPS orbital speeds will tick more slowly than stationary ground clocks by a factor of $\frac{v^2}{2c^2} \approx 10^{-10}$, or result in a delay of about 7 μs/day, where the orbital velocity is v = 4 km/s, and c = the speed of light. This time dilation effect has been measured and verified using the GPS.

The effect of gravitational frequency shift on the GPS due to general relativity is that a clock closer to a massive object will be slower than a clock farther away. Applied to the GPS, the receivers are much closer to Earth than the satellites, causing the GPS clocks to be faster by a factor of 5×10^{-10}, or about 45.9 μs/day. This gravitational frequency shift is noticeable.

When combining the time dilation and gravitational frequency shift, the discrepancy is about 38 microseconds per day, a difference of 4.465 parts in 10^{10}. Without correction, errors in the initial pseudorange of roughly 10 km/day would accumulate. This initial pseudorange error is corrected in the process of solving the navigation equations. In addition the elliptical, rather than perfectly circular, satellite orbits cause the time dilation and gravitational frequency shift effects to vary with time. This eccentricity effect causes the clock rate difference between a GPS satellite and a receiver to increase or decrease depending on the altitude of the satellite.

To compensate for the discrepancy, the frequency standard on board each satellite is given a rate offset prior to launch, making it run slightly slower than the desired frequency on Earth; specifically, at 10.22999999543 MHz instead of 10.23 MHz. Since the atomic clocks on board the GPS satellites are precisely tuned, it makes the system a practical engineering application of the scientific theory of relativity in a real-world environment. Placing atomic clocks on artificial satellites to test Einstein's general theory was proposed by Friedwardt Winterberg in 1955.

Calculation of Time Dilation

To calculate the amount of daily time dilation experienced by GPS satellites relative to Earth we need to separately determine the amounts due to special relativity (velocity) and general relativity (gravity) and add them together.

The amount due to velocity will be determined using the Lorentz transformation. This will be:

$$\frac{1}{\gamma} = \sqrt{1 - \frac{v^2}{c^2}}$$

For small values of v/c, by using binomial expansion this approximates to:

$$\frac{1}{\gamma} \approx 1 - \frac{v^2}{2c^2}$$

The GPS satellites move at 3874 m/s relative to Earth's center. We thus determine:

$$\frac{1}{\gamma} \approx 1 - \frac{3874^2}{2\left(2.998 \times 10^8\right)^2} \approx 1 - 8.349 \times 10^{-11}$$

This difference below 1 of 8.349×10^{-11} represents the fraction by which the satellites' clocks move slower than Earth's. It is then multiplied by the number of nanoseconds in a day:

$$-8.349 \times 10^{-11} \times 60 \times 60 \times 24 \times 10^9 \approx -7214 \text{ ns}$$

That is, the satellites' clocks lose 7,214 nanoseconds a day due to special relativity effects.

Note that this speed of 3874 m/s is measured relative to Earth's center rather than its surface where the GPS receivers (and users) are. This is because Earth's equipotential makes net time dilation equal across its geodesic surface. That is, the combination of Special and General effects make the net time dilation at the equator equal to that of the poles, which in turn are at rest relative to the center. Hence we use the center as a reference point to represent the entire surface.

The amount of dilation due to gravity will be determined using the gravitational time dilation equation:

$$\frac{1}{\gamma} = \sqrt{1 - \frac{2GM}{rc^2}}$$

For small values of M/r, by using binomial expansion this approximates to:

$$\frac{1}{\gamma} \approx 1 - \frac{GM}{rc^2}$$

We are again only interested in the fraction below 1, and in the difference between Earth and the satellites. To determine this difference we take:

$$\Delta\left(\frac{1}{\gamma}\right) \approx \frac{GM_{\text{earth}}}{R_{\text{earth}}c^2} - \frac{GM_{\text{earth}}}{R_{\text{gps}}c^2}$$

Earth has a radius of 6,357 km (at the poles) making R_{earth} = 6,357,000 m and the satellites have an altitude of 20,184 km making their orbit radius R_{gps} = 26,541,000 m. Substituting these in the above equation, with $M_{earth} = 5.974\times10^{24}$, $G = 6.674\times10^{-11}$, and $c = 2.998\times10^{8}$ (all in SI units), gives:

$$\Delta\left(\frac{1}{\gamma}\right) \approx 5.307\times10^{-10}$$

This represents the fraction by which the satellites' clocks move faster than Earth's. It is then multiplied by the number of nanoseconds in a day:

$$5.307\times10^{-10}\times60\times60\times24\times10^{9} \approx 45850 \text{ ns}$$

That is, the satellites' clocks gain 45,850 nanoseconds a day due to general relativity effects. These effects are added together to give (rounded to 10 ns):

$$45850 - 7210 = 38640 \text{ ns}$$

Hence the satellites' clocks gain approximately 38,640 nanoseconds a day or 38.6 μs per day due to relativity effects in total.

In order to compensate for this gain, a GPS clock's frequency needs to be slowed by the fraction:

$$5.307\times10^{-10} - 8.349\times10^{-11} = 4.472\times10^{-10}$$

This fraction is subtracted from 1 and multiplied by the pre-adjusted clock frequency of 10.23 MHz:

$$(1 - 4.472\times10^{-10}) \times 10.23 = 10.22999999543$$

That is, we need to slow the clocks down from 10.23 MHz to 10.22999999543 MHz in order to negate the effects of relativity.

Sagnac Distortion

GPS observation processing must also compensate for the Sagnac effect. The GPS time scale is defined in an inertial system but observations are processed in an Earth-centered, Earth-fixed (co-rotating) system, a system in which simultaneity is not uniquely defined. A coordinate transformation is thus applied to convert from the inertial system to the ECEF system. The resulting signal run time correction has opposite algebraic signs for satellites in the Eastern and Western celestial hemispheres. Ignoring this effect will produce an east–west error on the order of hundreds of nanoseconds, or tens of meters in position.

Natural Sources of Interference

Since GPS signals at terrestrial receivers tend to be relatively weak, natural radio signals or scattering of the GPS signals can desensitize the receiver, making acquiring and tracking the satellite signals difficult or impossible.

Space weather degrades GPS operation in two ways, direct interference by solar radio burst noise in the same frequency band or by scattering of the GPS radio signal in ionospheric irregularities referred to as scintillation. Both forms of degradation follow the 11 year solar cycle and are a maximum at sunspot maximum although they can occur at any time. Solar radio bursts are associated with solar flares and coronal mass ejections (CMEs) and their impact can affect reception over the half of the Earth facing the sun. Scintillation occurs most frequently at tropical latitudes where it is a night time phenomenon. It occurs less frequently at high latitudes or mid-latitudes where magnetic storms can lead to scintillation. In addition to producing scintillation, magnetic storms can produce strong ionospheric gradients that degrade the accuracy of SBAS systems.

Artificial Sources of Interference

In automotive GPS receivers, metallic features in windshields, such as defrosters, or car window tinting films can act as a Faraday cage, degrading reception just inside the car.

Man-made EMI (electromagnetic interference) can also disrupt or jam GPS signals. In one well-documented case it was impossible to receive GPS signals in the entire harbor of Moss Landing, California due to unintentional jamming caused by malfunctioning TV antenna preamplifiers. Intentional jamming is also possible. Generally, stronger signals can interfere with GPS receivers when they are within radio range or line of sight. In 2002 a detailed description of how to build a short-range GPS L1 C/A jammer was published in the online magazine Phrack.

The U.S. government believes that such jammers were used occasionally during the War in Afghanistan, and the U.S. military claims to have destroyed six GPS jammers during the Iraq War, including one that was destroyed with a GPS-guided bomb. A GPS jammer is relatively easy to detect and locate, making it an attractive target for anti-radiation missiles. The UK Ministry of Defence tested a jamming system in the UK's West Country on 7 and 8 June 2007.

Some countries allow the use of GPS repeaters to allow the reception of GPS signals indoors and in obscured locations; while in other countries these are prohibited as the retransmitted signals can cause multi-path interference to other GPS receivers that receive data from both GPS satellites and the repeater. In the UK Ofcom now permits the use of GPS/GNSS Repeaters under a ‹light licensing' regime.

Due to the potential for both natural and man-made noise, numerous techniques continue to be developed to deal with the interference. The first is to not rely on GPS as a sole source. "IFR pilots should have a fallback plan in case of a GPS malfunction". Receiver Autonomous Integrity Monitoring (RAIM) is a feature included in some receivers, designed to provide a warning to the user if jamming or another problem is detected. The U.S. military has also deployed since 2004 their Selective Availability/Anti-Spoofing Module (SAASM) in the Defense Advanced GPS Receiver (DAGR). In demonstration videos the DAGR was shown to detect jamming and maintain its lock on the encrypted GPS signals during interference which caused civilian receivers to lose lock.

Assisted GPS

Assisted GPS or Augmented GPS (abbreviated generally as A-GPS and less commonly as aGPS) is a system that often significantly improves the startup performance i.e., time-to-first-fix (TTFF) of a GPS satellite-based positioning system. A-GPS is extensively used with GPS-capable cellular phones, as its development was accelerated by the U.S. FCC's 911 requirement to make cell phone location data available to emergency call dispatchers.

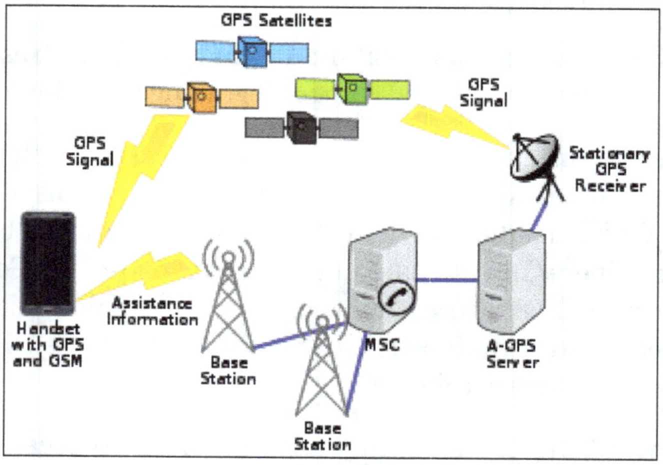

Assisted GPS.

Standalone/self-ruling GPS devices depend solely on information from satellites. A-GPS augments that by using cell tower data to enhance quality and precision when in poor satellite signal conditions. In exceptionally poor signal conditions, for example in urban areas, satellite signals may exhibit multipath propagation where signals skip off structures, or are weakened by meteorological conditions or tree canopy. Some standalone GPS navigators used in poor conditions can't fix a position because of satellite signal fracture and must wait for better satellite reception. A regular GPS unit may need as long as 12.5 minutes (the time needed to download the GPS almanac and ephemerides) to resolve the problem and be able to provide a correct location.

An assisted GPS system can address these problems by using external data. Utilizing this system can come at a cost to the user. For billing purposes, network providers often count this as a data access, which can cost money, depending on the plan.

To be precise, A-GPS features depend mostly on an Internet network or connection to an ISP (or CNP, in the case of CP/mobile-phone device linked to a cellular network provider data service). A mobile (cell phone, smart phone) device with just an L1 front-end radio receiver and no GPS acquisition, tracking, and positioning engine only works when it has an internet connection to an ISP/CNP, where the position fix is calculated offboard the device itself. It doesn't work in areas with no coverage or internet link (or nearby base transceiver station (BTS) towers, in the case on CNP service coverage area). Without any of those resources, it can't connect to the A-GPS servers usually provided by CNPs. On the other hand, a mobile device with a GPS chipset requires no data connection to capture and process GPS data into a position solution, since it receives data directly from the GPS satellites and is able to calculate a position fix itself. However, the availability of a data connection can provide assistance to improve the performance of the GPS chip on the mobile device.

Assistance falls into two categories:

- Mobile Station Based (MSB): Information used to acquire satellites more quickly.
    - It can supply orbital data or almanac for the GPS satellites to the GPS receiver, enabling the GPS receiver to lock to the satellites more rapidly in some cases.
    - The network can provide precise time.
- Mobile Station Assisted (MSA): Calculation of position by the server using information from the GPS receiver.
    - The device captures a snapshot of the GPS signal, with approximate time, for the server to later process into a position.
    - The assistance server has a good satellite signal and plentiful computation power, so it can compare fragmentary signals relayed to it.
    - Accurate, surveyed coordinates for the cell site towers allow better knowledge of local ionospheric conditions and other conditions affecting the GPS signal than the GPS receiver alone, enabling more precise calculation of position.

As an additional benefit, in mobile station assisted implementations, the amount of processing and software required for a GPS receiver can be reduced by offloading most of the work onto the assistance server.

A typical A-GPS-enabled receiver uses a data connection (Internet or other) to contact the assistance server for aGPS information. If it also has functioning autonomous GPS, it may use standalone GPS, which is sometimes slower on time to first fix, but does not depend on the network, and therefore can work beyond network range and without incurring data-usage fees. Some A-GPS devices do not have the option of falling back to standalone or autonomous GPS.

Many mobile phones combine A-GPS and other location services, including Wi-Fi positioning system and cell-site multilateration and sometimes a hybrid positioning system.

High-Sensitivity GPS is an allied technology that addresses some of these issues in a way that does not require additional infrastructure. However, unlike some forms of A-GPS, high-sensitivity GPS cannot provide a fix instantaneously when the GPS receiver has been off for some time.

Standalone GPS provides first position in approximately 30–40 seconds. A standalone GPS needs orbital information of the satellites to calculate the current position. The data rate of the satellite signal is only 50 bit/s, so downloading orbital information like ephemerides and the almanac directly from satellites typically takes a long time, and if the satellite signals are lost during the acquisition of this information, it is discarded and the standalone system has to start from scratch. In A-GPS, the network operator deploys an A-GPS server, a cache server for GPS data. These A-GPS servers download the orbital information from the satellite and store it in the database. An A-GPS-capable device can connect to these servers and download this information using mobile-network radio bearers such as GSM, CDMA, WCDMA, LTE or even using other radio bearers such as Wi-Fi or LoRa. Usually the data rate of these bearers is high; hence downloading orbital information takes less time.

Modes of Operation

Mobile Station Assisted

In MSA mode A-GPS operation, the A-GPS capable device receives acquisition assistance, reference time and other optional assistance data from a mobile service provider. The mobile service provider continuously logs GPS information (mainly the almanac) from the GPS satellites using an A-GPS server in its system. With the help of the above data (the data received from the mobile device and the data already present in an A-GPS server) the A-GPS server calculates the position and sends it back to the A-GPS device.

Mobile Station Based

In MSB mode A-GPS operation, the A-GPS device receives ephemeris, reference location, reference time and other optional assistance data from the A-GPS server. With the help of the above data, the A-GPS device receives signals from the visible satellites and calculates the position.

Standards

A-GPS protocols are part of Positioning Protocol defined by two different standardization bodies, 3GPP and Open Mobile Alliance (OMA).

Control Plane Protocol

Defined by the 3GPP for various generations of mobile phone systems. These protocols are defined for circuit switched networks. The following positioning protocols have been defined.

- RRLP: 3GPP defined RRLP (Radio Resource Location Protocol) to support positioning protocol on GSM networks.
- TIA 801: CDMA2000 family defined this protocol for CDMA 2000 networks.
- RRC position protocol: 3GPP defined this protocol as part of the RRC standard for UMTS network.
- LPP: 3GPP defined LPP or LTE positioning protocol for LTE networks.

User Plane Protocol

Defined by the OMA to support positioning protocols in packet switched networks. Two generations of User Plane Protocol have been defined.

- SUPL V1.0.
- SUPL V2.0.

Differential GPS

A Differential Global Positioning System (DGPS) is an enhancement to the Global Positioning System (GPS) which provides improved location accuracy, in the range of operations of each system, from the 15-meter nominal GPS accuracy to about 1-3 cm in case of the best implementations.

Each DGPS uses a network of fixed ground-based reference stations to broadcast the difference between the positions indicated by the GPS satellite system and known fixed positions. These stations broadcast the difference between the measured satellite pseudoranges and actual (internally computed) pseudoranges, and receiver stations may correct their pseudoranges by the same amount. The digital correction signal is typically broadcast locally over ground-based transmitters of shorter range.

The United States Coast Guard (USCG) and the Canadian Coast Guard (CCG) each run DGPSes in the United States and Canada on longwave radio frequencies between 285 kHz and 325 kHz near major waterways and harbors. The USCG's DGPS was named NDGPS (Nationwide DGPS) and was jointly administered by the Coast Guard and the U.S. Department of Defense's Army Corps of Engineers (USACE). It consisted of broadcast sites located throughout the inland and coastal portions of the United States including Alaska, Hawaii and Puerto Rico. Other countries have their own DGPS.

A similar system which transmits corrections from orbiting satellites instead of ground-based transmitters is called a Wide-Area DGPS (WADGPS) or Satellite Based Augmentation System.

Transportable DGPS reference station *Baseline HD* by CLAAS for use in satellite-assisted steering systems in modern agriculture.

Operation

DGPS Reference Station (choke ring antenna).

A reference station calculates differential corrections for its own location and time.

Users may be up to 200 nautical miles (370 km) from the station, however, and some of the compensated errors vary with space: specifically, satellite ephemeris errors and those introduced by ionospheric and tropospheric distortions. For this reason, the accuracy of DGPS decreases with distance from the reference station. The problem can be aggravated if the user and the station lack "inter visibility"—when they are unable to see the same satellites.

Post Processing

Post-processing is used in Differential GPS to obtain precise positions of unknown points by relating them to known points such as survey markers.

The GPS measurements are usually stored in computer memory in the GPS receivers, and are subsequently transferred to a computer running the GPS post-processing software. The software computes baselines using simultaneous measurement data from two or more GPS receivers.

The baselines represent a three-dimensional line drawn between the two points occupied by each pair of GPS antennas. The post-processed measurements allow more precise positioning, because most GPS errors affect each receiver nearly equally, and therefore can be cancelled out in the calculations.

Differential GPS measurements can also be computed in real time by some GPS receivers if they receive a correction signal using a separate radio receiver, for example in Real Time Kinematic (RTK) surveying or navigation.

The improvement of GPS positioning doesn't require simultaneous measurements of two or more receivers in any case, but can also be done by special use of a *single* device. In the 1990s when even handheld receivers were quite expensive, some methods of quasi-differential GPS were developed, using the receiver by quick turns of positions or loops of 3-10 survey points.

StarFire

StarFire is a wide-area differential GPS developed by John Deere's NavCom and precision farming groups. StarFire broadcasts additional "correction information" over satellite L-band frequencies around the world, allowing a StarFire-equipped receiver to produce position measurements accurate to well under one meter, with typical accuracy over a 24-hour period being under 4.5 cm. StarFire is similar to the FAA's differential GPS Wide Area Augmentation System (WAAS), but considerably more accurate due to a number of techniques that improve its receiver-end processing.

Addressing GPS Inaccuracy

In theory the GPS signal with Selective Availability turned off offers accuracy on the order of 3 m. In practice, typical accuracy is about 15 m.

Of this 12 m, about 5 m is due to distortion from "billows" in the ionosphere, which introduce propagation delays that makes the satellite appear farther away than it really is. Another 3 to 4 m is accounted for by errors in the satellite ephemeris data, which is used to calculate the positions of the GPS satellites, and by clock drift in the satellite's internal atomic clocks.

dGPS correct for these errors by comparing the position measured using GPS with a known highly accurate ground reference, and then calculating the difference and broadcasting it to users. Some of these corrections apply to any location the corrections to the clocks and ephemeris data for instance. In contrast, the billows cover only a certain portion of the sky, so a correction measured at any one ground station is only really useful for receivers located nearby. To make the corrections accurate over a large area, one would need to deploy many ground reference stations and broadcast a considerable amount of data for finely divided locations. For instance, WAAS uses twenty-five stations in the continental US, developing a grid spaced 5x5 degrees.

StarFire instead uses an advanced receiver to correct for ionospheric effects internally. To do this, it captures the P(Y) signal that is broadcast on two frequencies, L1 and L2, and compares the effects of the ionosphere on the propagation time of the two. Using this information, the ionospheric effects can be calculated to a very high degree of accuracy, meaning the StarFire dGPS can compensate for variations in propagation delay. The second P(Y) signal is encrypted and cannot be used by civilian receivers directly, but StarFire doesn't use the data contained in the signal; it only compares the phase of the two signals instead. This is expensive in terms of electronics, requiring a second tuner and excellent signal stability to be useful, which is why the StarFire-like solution is not more widely used (at least when it was being created).

With the ionospheric correction handled internally, the StarFire dGPS signal is greatly reduced in the amount of information it needs to carry, which consists of a set of correction signals for the satellite data alone. Since these corrections are globally valid, and there are only 24 satellites in operation at any time, the total amount of information is quite limited. StarFire broadcasts this data at 300 bits per second, repeating once a second. The corrections are generally valid for about 20 minutes. In addition to ephemeris and clock corrections, the signal also contains information on the health of each satellite, offering quality-of-service data in near real-time, with about a 3-second delay in updating the signals from the ground station.

Versions

StarFire has developed through two versions. The first, retroactively known as SF1, offered 1-sigma accuracy of about 1 m. Its error was about 15 to 30 cm, meaning that while the displayed position (absolute accuracy) might be off by about 1 m, it could return you to within centimeters of a previously measured spot (relative accuracy). This was enough for the intended role, field surveying. This system was first offered in 1998, and since its replacement the SF1 signal is apparently now offered for free.

Survey crew using the pole-mounted Navcom SF-2040G receiver.

The newer system, SF2, was introduced in 2004. It dramatically improves accuracy, with a 1-sigma absolute accuracy of about 4.5 cm. In other words, StarFire will leave you within 4.5 cm of a particular geographical point 65% of the time, and be accurate to under 10 cm around 95% of the time (2-sigma). The relative accuracy is likewise improved, to about 2.5 cm.

Even if the StarFire correction signal is lost for more than 20 minutes, the internal ionospheric corrections alone result in accuracy of about 3 m. StarFire receivers also receive WAAS signals, ignoring their ionospheric data and using their (less detailed) ephemeris and clock adjustment data to provide about 50 cm accuracy. In comparison, "normal" GPS receivers generally offer about 15 m accuracy, and ones using WAAS improve this to about 3 m.

Reference Stations

When initially deployed, StarFire used seven reference stations in the continental US. The corrections generated at these stations are sent to two redundant processing stations (one co-located with a reference/monitor site), and then the resulting signal is uplinked from an east-coast US station. All of the stations are linked over the internet, with dedicated ISDN lines and VSAT links as backups. The resulting signals were broadcast from an Inmarsat III channel.

Additional StarFire networks were later set up in South America, Australia and Europe, each run from their own reference stations and sending data to their own satellites. As use of the system grew, the decision was made to link the various "local area" networks into a single global one. Today the StarFire network uses twenty-five stations worldwide, calculating and uplinking data from the US stations as before. The data collected at these stations is not location-dependent, in contrast to most dGPS, and the large number of sites is used primarily for redundancy.

Variants

John Deere also sells a Real Time Kinematic dGPS, StarFire RTK. RTK consists of a small tripod-mounted GPS receiver that uses StarFire signals to perform its own dGPS calculations relative to a point, normally the corner of a field. The unit then broadcasts these corrections over a radio link to the equipment-mounted receivers. RTK offers absolute accuracy of about 2 cm, and relative accuracy in the millimeters. This sort of accuracy is used for fully automated equipment with autodrive systems.

GPS Applications

GPS has in the last years been intensively used as receivers for directions and as navigations app that tells us where we are and how to move between points. They have also been used to make digital maps. However, GPS apps are not only used by civilians but also by different organizations for different purposes. For example, it is useful to pilots, surveyors, farmers, boat captains, military and scientists among others. Apart from the usually GPS that we use in our phones, other highly advanced apps are incredibly accurate and are used for specific tasks in different industries. Some of the industries that use this application include:

- Aviation: Most of the modern aircraft use GPS receivers to provide the pilots and passenger with real-time aircraft position. They also provide a map of various destinations depending on where the aircraft operates. This application is also used by the airline operators to decide which route is the fastest, safest, and most fuel-efficient among the destinations. They also use the app to track the aircraft and direct the pilot in the case where there is a change in the weather conditions or any other issue that may arise.

- Marine: Highly accurate navigation app is needed by boat captains to enable them to navigate through waters to their destinations. These applications ensure that the channels are clear and there are no obstacles that can hinder their navigation. They are also required in the marine departments since they are used to map and position dredging operations in rivers, sandbars, and wharves to ensure that other boats are aware of how deep they should get.

- Farming: Farmers have a specific season for planting, weeding, and harvesting, and due to the repeat in the seasons, they put the GPS receiver on their tractors and other farming equipment. This allows them to map their plantations and ensure that they return to precisely the same time when planting or weeding in the next season. This strategy is effective especially in seasons when it's foggy with less visibility since the machine will still operate since its GPS and not visual reference guides it. More so, its high accuracy makes it suitable for use in mapping soil sample locations, and the farmers can locate the areas that have soils suitable for farming.

- Science: This is one field that intensively uses the GPS app especially since there are numerous braches and departments in the science field and this include physics, biology and earth science to mention a few. Before the invention of this app, scientists used metal and plastic bands to tag animals, and they would follow them to various locations while monitoring them. However, since the invention and enhancement of the navigation app, it has helped scientist to fit the animal with the collars and the app can automatically record the animal's movement and the information is transmitted to a through a satellite to the researcher. This means that they can trace the location of the animals' movement without having to relocate them physically. Most of the sciences taught in schools will have more practical evidence especially since GPS provides accurate data. Earth scientist also uses the app to study how landscapes change over time. Or any geographic area that they may be interested.

- Surveying: Surveying is one of the uses of GPS that are essential especially since it is used in mapping and measuring various measures on the earth surface and underwater. It is used in determining land boundaries, mapping sea floors, and highlighting the changes in the shape of structures. The best thing about high accuracy GPS app is that the surveyors can set it up over a single point and establish a reference marker. They can also use it in a moving configuration to map the boundaries of particular features. With the data obtained from the application, they can easily key in the details into software that will help them offer their customers with a detailed chart.

- Military: The US Department of Defense was the first to develop the GPS app system, and since then the system has been adopted by numerous military forces around the world. Other countries have even decided to develop their satellite navigation networks as a defense mechanism during war times. Today, there has been a diverse use of the app, and it can be used to map the location of vehicles and other machinery such as missiles during a war. This is a technique used purposely to protect the soldiers and also manage resources.

- Financial Services: Financial organizations such as banks use this app to schedule and determine local and international money transfers. They are also using it to provide audit trails of financial transactions. More so, since more than 80% of the transactions are made through debit and credit cards, it has been easier to provide a higher level of timing accuracy. The GPS satellite is necessary for the financial field since it allows for data and time stamps of Electronic Funds Transfers.

- Telecommunications: Telecommunications especially the mobile telephones use this app to provide its users with accuracy, reliability, and stability of their operations. Although other clocks can provide this, the GPS supports the derivation of synchronized time zones with the help of the satellite signals.

- Heavy Vehicle Guidance: Heavy tack machines used in mining and constructions also use this technology. For example in highway construction, the marker pegs and surveyors have been replaced by the improvised in-cabin vehicle guidance and control systems. This makes the work easier since the driver only needs to follow the surveyor's pre-programmed site plan. Thus, with this app, the constructors can achieve close tolerance for position, gradient, and level.
- Road Transportation: Majority of users of this technology are taxi services, emergency vehicle location, commercial fleet management and freight tracking, public transport monitoring, dispatch, and navigation. Private car owners also use the app, and most of the new car models come with a factory-fitted GPS.
- Social Activities: Some of the social activities that have incorporated the use of this technology include cross-country cycling, skiing, hiking skydiving, paragliding, geotagging photographs, geocaching, geodashing among others.
- Locating Positions: GPS navigation system is commonly used to designate positions. It is used by travelers, researchers and the everyday citizens. More so, you can use it to locate websites that offer certain services in your country. Certain websites block access by people from a specific location. For example, if you are looking for a website that offers academic assistance in your area such as online custom writing service, then as long as your GPS location is allowed into the website, you will get a list of recommendable website.
- Tracking for Law Enforcement: The police also use it during investigations to catch suspects and criminals. They also use it to keep track of people who feel that their lives may be in danger.

- Locating your Pets: Pets such as cats and dogs can easily get distracted and go missing. A GPS microchip will help you locate your pet's whereabouts.
- Keeping Watch of the Elderly: Some elderly people don't like when someone monitors them and they at times wander alone in unfamiliar places and may find it difficult to trace their way back home. Or maybe one may fall down and may need emergency medical assistance. This device can be used where, if they need any help, they only have to press a button, and the caregiver will be alerted.

They come in tiny units that can be used on their clothes. Some are also wearable watches that are not conspicuous.

- Seeking Treasure: Uses of GPS are intensive, and tech-geeks also use the technology when hunting treasure with friends. The clues are hidden in sites programmed on treasure maps, and these apps are uploaded to a particular website where people join and compete for the prizes.
- Securing Artworks: Some paintings worth millions and are susceptible to theft. Although some are insured, owners also use trackers as additional security to help them locate them when stolen.
- Hunting and Fishing: Using this technology allows you to mark the spots that you would want to return to your next hunting or fishing trip.
- Easy Access to Emergency Roadside Support: In case of an accident or an emergency, you can seek assistance using the pre-programmed emergency numbers on your smartphone. The best thing about using this app is that the emergency crew can trace your current location without having to provide any details.
- GPS Technology Combined with Intelligence Vehicle Highway Systems is used to Improve Highway Safety and Ease Congestion: Integrated Traffic Management Systems requires accurate data that is available in real time. This is to ensure that the necessary authorities can manage the travel time and speed to monitor and control congestion. Thus, the GPS is used to collect the required data, and it has been proved to be 50% more efficient in terms of manpower.

- Public Safety and Disaster Relief: The best thing about GPS is that it can be used in any weather or environmental condition which is the reason why is preferred for use during disaster management. The emergency vehicles and supplies are tracked using GPS.
- In the Robotic field: Just like GPS provides direction to human, the same case applies to robots. GPS enhances navigation of mobile robots and makes them applicable to diverse fields such as outdoor industrial work, mapping, and agriculture.
- Can be used by Disabled People: People with special needs are at times left on their one when their caretakers and loved ones have to work. The GPS tracker is not only used to track their location, but it can be of great importance in the case of emergencies.

School Bus Tracking: School Bus GPS tracking system is used in gathering information about which route is most efficient. It also provides the bus drivers with real-time directions whenever they get lost. School management can use this technology to ensure that the bus drivers follow the authorized routes and that they do not take unnecessary breaks during working hours.

- Trailer Tracking: GPS trailer tracking system is an important aspect that improves security and increases efficiency as well as reduces costs. The management uses the GPS app tracking system to monitor their fleet and equipment. It also provides accurate and detailed reporting which gives the owner peace of mind.
- Boat Tracking: Boat tracking system protects your boat from being stolen. Installing a navigation app allows you to see the location of your boat in real time. More so, there are systems that enable you to see its fuel consumption. This technology not only gives you peace of mind but it also allows you to effectively schedule the hours of delivery and the route that is fuel efficient.
- Business Tracking: These tracking systems have been used by various businesses ranging from small, medium and large organizations to remain competitive.

The technology provides business owners with real-time information about their business and other relevant information. The devices vary depending on the business, but there is a whole lot of advantages that a company can obtain from using a business tracking system.

- Live Recording: Live streaming video and tracking are other uses of GPS app, and the best thing about it is that it offers real-time data and you can record activities as they happen and through the transmission, through the satellite, you can view the happenings million miles away.
- Speed Monitoring: GPS tracker provides accurate speed estimates of vehicles or fleet which are transmitted to a central hub where they can read in real time and recorded for future reference. Keeping track of speed allows you to reduce the cost of fuel consumption, a lesser number of speed tickets, reduces the rate of accidents, and minimizes wear and tear on your vehicle.
- Graphics Reporting: The navigation app is useful in providing a graphic report of the occurrences as they happen. This could either be the distance covered or the fuel consumed.
- Geo-mapping and Landmarking: This system provides integrated and accurate landmarking and mapping information. It is known to be highly accurate and faster than any conventional strategy. There is minimal labor required and less amount of equipment to be used.
- Making Electronic and Rough Topographic Maps: This app is effective in creating coordinates for contour and topographic maps. The best thing about using GPS is that it is accurate, convenient and can be used in any location without requiring much supervision.

References

- Global-Positioning-System: newworldencyclopedia.org, Retrieved 10 January, 2020
- Penttinen, Jyrki T. J. The Telecommunications Handbook: Engineering Guidelines for Fixed, Mobile and Satellite Systems. John Wiley & Sons. ISBN 9781119944881
- Grewal, Mohinder S.; Weill, Lawrence Randolph; Andrews, Angus P. (2001). Global positioning systems, inertial navigation, and integration. John Wiley and Sons. ISBN 978-0-47135-032-3
- McNamara, Joel (2008), GPS for Dummies (2nd ed.), ISBN 978-0-470-15623-0
- 50-uses-or-applications-of-gps: grindgis.com, Retrieved 19 May, 2020

4

Global Positioning System Based Software

There are many GPS based software that are used for navigation and tracking purposes. Some of the software are Google Earth, Google Maps, Google Latitude, GPS Logger, etc. The topics elaborated in this chapter will help in gaining a better perspective of these different GPS based software.

Google Earth

Google Earth is a geobrowser that accesses satellite and aerial imagery, ocean bathymetry, and other geographic data over the internet to represent the Earth as a three-dimensional globe. Geobrowsers are alternatively known as virtual globes or Earth browsers. Google also refers to Google Earth as a "geographic browser". Other examples of geobrowsers are NASA's World Wind, ESRI's ArcGIS Explorer, GeoFusions's GeoPlayer, and EarthBrowser by Lunar Software. Google Earth is available on the web for free as well for purchase in more advanced versions. While the free version offers numerous features that are useful in educational settings, the Pro version offers additional capabilities such as higher resolution printing and saving of images and the ability to open ESRI shapefiles. The free version of Google Earth as well as Pro are available through Google's Explore, Search, and Discover page.

Three Versions of Google Earth

- Free: Intended for home and personal use, this product has many features, including displaying satellite and aerial imagery, a growing set of layers of mappable data, the ability to display third party data, tools for creating new data, and the ability to import GPS data. Schools may use the free version of Google Earth, and Google has created a Geo Education site to provide helpful information on using Google Earth, Maps, Sky, and SketchUp in the K-12 classroom. Higher education institutions may also install the free version for non-commercial use.

- Pro: This version, developed for commercial use, adds movie making, as well as importing ESRI shapefiles and MapInfo tab files, can measure areas of circles and polygons, and can print and save high-resolution images.
- Enterprise: This product makes imagery and other geospatial data available to employees within organizations such as corporations.

Each of these versions of Google Earth can be used to read and create data in KML (Keyhole Markup Language) format, which enables educators, students, and other users to share data.

Google Earth provides search capabilities and the ability to pan, zoom, rotate, and tilt the view of the Earth. It also offers tools for creating new data and a growing set of layers of data, such as volcanoes and terrain that reside on Google's servers, and can be displayed in the view.

It also uses elevation data primarily from NASA's Shuttle Radar Topography Mission (SRTM) to offer a terrain layer, which can visualize the landscape in 3D. For some locations, such as most of the western portion of the United States, the terrain data is provided at significantly higher resolutions.

Google Earth is not a Geographic Information System (GIS) with the extensive analytical capabilities of ArcGIS or MapInfo, but is much easier to use than these software packages.

It is available for several operating systems, namely:

- Microsoft Windows 2000.
- Microsoft Windows XP.
- Microsoft Windows Vista.
- Mac OS X version 10.3.9 or higher.
- Linux.
- Free BSD.

Google Maps and the Google Earth API are products that include some of the features of Google Earth, and can used to embed interactive maps into web pages. The Google Maps site was originally created as a service for providing driving directions. But curious enthusiasts examined the JavaScript code that supports the service, and quickly learned to create their own customized Google Maps. Subsequently, Google made the details of the Google Maps API public for all to use for creating customized user interfaces. In contrast, the Google Earth API was expressly created to enable the public to create customized three-dimensional map interfaces in web pages that have the look and feel of Google Earth. Both Google Maps and the Google Earth API offer the same imagery that is available through Google Earth, along with the capability of adding features, such as customized buttons, forms, and panels that can work with and display data. To create these specialized applications, developers use AJAX, which combines writing code in JavaScript with accessing data that is stored in XML format. In fact, KML is actually a form of XML. Google Maps can be used in all major web browsers and platforms; however, the Google Earth API is currently available for Windows and the Macintosh, but not Linux. It is restricted to a subset of browsers, and requires the user to download a plugin. Another mapping technology available through Google is the Google Mapplets API. Mapplets are small applications that run within Google Maps rather than being hosted on a third-party web site.

Coding in JavaScript is not needed in order to create simple Google Maps. Setting up simple customized Google Maps and posting them on the web requires logging into a free Google account and activating the My Maps tab. A set of tools enables the creation of points, lines and polygons on the map. The map is assigned a url that can be used to share it students or others. HTML code is also displayed that can be used to embed the map into a web page. A web browser is sufficient for viewing maps created in this manner. A KML file that has been stored on a web server can be opened in Google Maps by entering its url in the Google Maps search box. While easy to create and share, these simple Google Maps are not as feature-rich as Google Earth.

There is even a version of Google Earth for the iPhone. This product can display the same imagery that is available on the desktop versions, perform searches, and link to Wikipedia articles about places of interest, but it does not have all the features of the desktop versions of Google Earth.

Google Maps

Google Maps is a web mapping service developed by Google. It offers satellite imagery, aerial photography, street maps, 360° panoramic views of streets (Street View), real-time traffic conditions, and route planning for traveling by foot, car, bicycle and air (in beta), or public transportation.

Google Maps began as a C++ desktop program at Where 2 Technologies. In October 2004, the company was acquired by Google, which converted it into a web application. After additional acquisitions of a geospatial data visualization company and a realtime traffic analyzer, Google Maps was launched in February 2005. The service's front end utilizes JavaScript, XML, and Ajax. Google Maps offers an API that allows maps to be embedded on third-party websites, and offers a locator for businesses and other organizations in numerous countries around the world. Google Map Maker allowed users to collaboratively expand and update the service's mapping worldwide but was discontinued from March 2017. However, crowdsourced contributions to Google Maps were not discontinued as the company announced those features would be transferred to the Google Local Guides program.

Google Maps' satellite view is a "top-down" or bird's-eye view; most of the high-resolution imagery of cities is aerial photography taken from aircraft flying at 800 to 1,500 feet (240 to 460 m), while most other imagery is from satellites. Much of the available satellite imagery is no more than three years old and is updated on a regular basis. Google Maps used a variant of the Mercator projection, and therefore could not accurately show areas around the poles. However, in August 2018, the desktop version of Google Maps was updated to show a 3D globe as the default projection. It is still possible to switch back to the Mercator Projection in the settings.

Google Maps for Android and iOS devices was released in September 2008 and features GPS turn-by-turn navigation along with dedicated parking assistance features. In August 2013, it was determined to be the world's most popular app for smartphones, with over 54% of global smartphone owners using it at least once. In 2012, Google reported having over 7,100 employees and contractors directly working in mapping. In 2017, along with several other services including YouTube, Chrome, Gmail, Search, and Google Play, Google Maps reached over 1 billion monthly users.

Functionality

Directions and Transit

Google Maps provides a route planner, allowing users to find available directions through driving, public transportation, walking, or biking. Google has partnered globally with over 800 public transportation providers to adopt General Transit Feed Specification (GTFS), making the data available to third parties. The app also has starred lines now to indicate users' transit route in the latest update in October 2019. Along with this the incognito mode, eyes-free walking navigation features have been released earlier.

Traffic Conditions

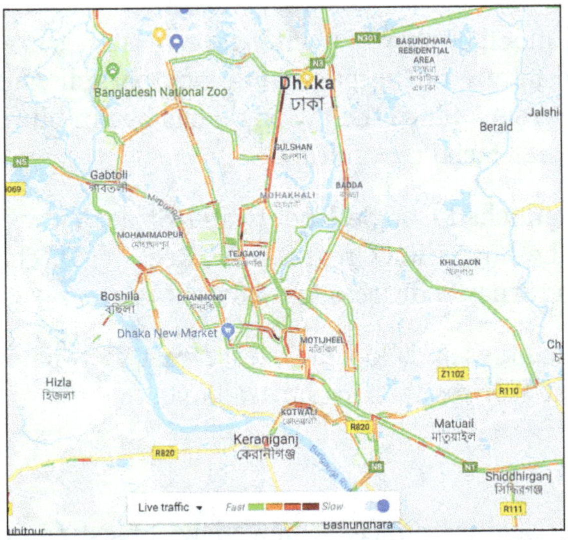

Screenshot of Google Maps with traffic option enabled.

In 2007, Google began offering traffic data as a colored overlay on top of roads and motorways to represent the speed of traffic. Crowdsourcing is used to obtain the GPS-determined locations of a large number of cellphone users, from which live traffic maps are produced.

Google has stated that the speed and location information it collects to calculate traffic conditions are anonymous. Options available in each phone's settings allow users not to share information about their location with Google Maps. Google stated, "Once you disable or opt out of My Location, Maps will not continue to send radio information back to Google servers to determine your handset's approximate location".

Street View

Google Maps car.

On May 25, 2007, Google released Google Street View, a new feature of Google Maps which provides 360° panoramic street-level views of various locations. On the date of release, the feature only included five cities in the US. It has since expanded to thousands of locations around the world. In July 2009, Google began mapping college campuses and surrounding paths and trails.

Street View garnered much controversy after its release because of privacy concerns about the uncensored nature of the panoramic photographs, although the views are only taken on public streets. Since then, Google has begun blurring faces and license plates through automated facial recognition.

In late 2014, Google launched Google Underwater Street View, including 2,300 kilometres (1,400 mi) of the Australian Great Barrier Reef in 3D. The images are taken by special cameras which turn 360 degrees and take shots every 3 seconds.

In 2017, in both Google Maps and Google Earth, Street View navigation of the International Space Station interior spaces became available.

45° Imagery

In December 2009, Google introduced a new view consisting of 45° angle aerial imagery, offering a "bird's-eye view" of cities. The first cities available were San Jose and San Diego. This feature was initially available only to developers via the Google Maps API. In February 2010, it was introduced as an experimental feature in Google Maps Labs. In July 2010, 45° imagery was made available in Google Maps in select cities in the United States and worldwide.

An example of the Leaning Tower of Pisa in the 45° view.

Business Listings

Google collates business listings from multiple on-line and off-line sources. To reduce duplication in the index, Google's algorithm combines listings automatically based on address, phone number, or geocode, but sometimes information for separate businesses will be inadvertently merged with each other, resulting in listings inaccurately incorporating elements from multiple businesses. Google allows business owners to verify their own business data through *Google My Business*, and has also recruited volunteers to check and correct ground truth data.

Google Maps can easily be manipulated by businesses which are not physically located in the area in which they record a listing. There are cases of people abusing Google Maps to overtake their competition by placing unverified listings on online directory sites knowing the information will roll across to Google (duplicate sites). The people that update these listings do not use a registered business name. Keywords and location details are placed on their Google Maps business title which overtake credible business listings. In Australia in particular, genuine companies and businesses are noticing a trend of fake business listings in a variety of industries.

Indoor Maps

In March 2011, indoor maps were added to Google Maps, giving users the ability to navigate themselves within buildings such as airports, museums, shopping malls, big-box stores, universities, transit stations, and other public spaces (including underground facilities). Google encourages owners of public facilities to submit floor plans of their buildings in order to add them to the service. Map users can view different floors of a building or subway station by clicking on a level selector that is displayed near any structures which are mapped on multiple levels.

My Maps

Google My Maps.

My Maps is a feature in Google Maps launched in April 2007 that enables users to create custom maps for personal use or sharing. Users can add points, lines, shapes, notes and images on top of Google Maps using a WYSIWYG editor. An Android app for My Maps, initially released in March 2013 under the name Google Maps Engine Lite, is also available.

Google Local Guides

Google Local Guides is a program launched by Google Maps to enable its users to contribute to Google Maps and provide them additional perks and benefits for the work. The program is partially a successor to Google Map Maker as features from the former program became integrated into the website and app.

The program consists of adding reviews, photos, basic information, videos and correcting information such as wheelchair accessibility.

Dark Mode

The Maps app also has a theme for navigation mode for users in line with current trends.

Implementation

A split view screenshot of Google Maps. In the bottom half the *Street Maps* is shown, while in the top half the *Street View* is shown. A user can zoom-in and out either of them independently of the zoom level of each.

As the user drags the map, the grid squares are downloaded from the server and inserted into the page. When a user searches for a business, the results are downloaded in the background for insertion into the side panel and map; the page is not reloaded. A hidden iframe with form submission is used because it preserves browser history. Like many other Google web applications, Google Maps uses JavaScript extensively. The site also uses JSON for data transfer rather than XML, for performance reasons.

The version of Google Street View for classic Google Maps required Adobe Flash. In October 2011, Google announced MapsGL, a WebGL version of Maps with better renderings and smoother transitions. Indoor maps uses JPG,.PNG,.PDF,.BMP, or.GIF, for floor plans.

Users who are logged into a Google Account can save locations so that they are overlaid on the map with various colored "pins" whenever they browse the application. These "Saved places" can be organised into user named lists and shared with other users. One default list "Starred places" also automatically creates a record in another Google product, Google Bookmarks.

Map Data and Imagery

The Google Maps terms and conditions state that usage of material from Google Maps is regulated by Google Terms of Service and some additional restrictions. Google has either purchased local map data from established companies, or has entered into lease agreements to use copyrighted map data. The owner of the copyright is listed at the bottom of zoomed maps. For example, street maps in Japan are leased from Zenrin. Street maps in China are leased from AutoNavi. Russian street maps are leased from Geocentre Consulting and Tele Atlas. Data for North Korea is sourced from the companion project Google Map Maker.

Street map overlays, in some areas, may not match up precisely with the corresponding satellite images. The street data may be entirely erroneous, or simply out of date: "The biggest challenge is the currency of data, the authenticity of data," said Google Earth representative Brian McClendon. As a result, in March 2008 Google added a feature to edit the locations of houses and businesses.

Restrictions have been placed on Google Maps through the apparent censoring of locations deemed potential security threats. In some cases the area of redaction is for specific buildings, but in other cases, such as Washington, D.C., the restriction is to use outdated imagery.

Google Maps API

After the success of reverse-engineered mashups such as chicagocrime.org and housingmaps.com, Google launched the Google Maps API in June 2005 to allow developers to integrate Google Maps into their websites. It was a free service that didn't require an API key until June 2018, when it was announced that an API key linked to a Google Cloud account with billing enabled would be required to access the API. The API currently does not contain ads, but Google states in their terms of use that they reserve the right to display ads in the future.

By using the Google Maps API, it is possible to embed Google Maps into an external website, on to which site-specific data can be overlaid. Although initially only a JavaScript API, the Maps API was expanded to include an API for Adobe Flash applications (but this has been deprecated), a service for retrieving static map images, and web services for performing geocoding, generating driving directions, and obtaining elevation profiles. Over 1,000,000 web sites use the Google Maps API, making it the most heavily used web application development API. In September 2011, Google announced it would deprecate the Google Maps API for Flash.

The Google Maps API is free for commercial use, provided that the site on which it is being used is publicly accessible and does not charge for access, and is not generating more than 25,000 map accesses a day. Sites that do not meet these requirements can purchase the Google Maps API for Business.

As of 21 June 2018, Google increased the prices of the Maps API and requires a billing profile.

Google Maps in China

Due to restrictions on geographic data in China, Google Maps must partner with a Chinese digital map provider in order to legally show Chinese map data. Since 2006, this partner has been AutoNavi.

Within China, the State Council mandates that all maps of China use the GCJ-02 coordinate system, which is offset from the WGS-84 system used in most of the world. google.*cn*/maps (formerly Google Ditu) uses the GCJ-02 system for both its street maps and satellite imagery. google.*com*/maps also uses GCJ-02 data for the street map, but uses WGS-84 coordinates for satellite imagery, causing the so-called China GPS shift problem.

Frontier alignments also present some differences between google.*cn*/maps and google.*com*/maps. On the latter, sections of the Chinese border with India and Pakistan are shown with dotted lines, indicating areas or frontiers in dispute. However, google.*cn* shows the Chinese frontier strictly according to Chinese claims with no dotted lines indicating the border with India and Pakistan. For example, the South Tibet region claimed by China but administered by India as a large part of Arunachal Pradesh is shown inside the Chinese frontier by google.*cn*, with Indian highways ending abruptly at the Chinese claim line. Google.*cn* also shows Taiwan and the South China Sea Islands as part of China. Google Ditu's street map coverage of Taiwan no longer omits major state organs, such as the Presidential Palace, the five Yuans, and the Supreme Court.

Feature-wise, google.*cn*/maps does not feature My Maps. On the other hand, while google.*cn* displays virtually all text in Chinese, google.*com*/maps displays most text (user-selectable real text as well as those on map) in English. This behavior of displaying English text is not consistent but intermittent sometimes it is in English, sometimes it is in Chinese. The criteria for choosing which language is displayed are not known publicly.

Discontinued Features

Google Latitude

Google Latitude was a feature from Google that lets users share their physical locations with other people. This service was based on Google Maps, specifically on mobile devices. There was an iGoogle widget for desktops and laptops as well. Some concerns were expressed about the privacy issues raised by the use of the service. On August 9, 2013, this service was discontinued, and in March 22, 2017, Google incorporated the features from Latitude into the Google Maps app.

Google Map Maker

In areas where Google Map Maker was available, for example, much of Asia, Africa, Latin America and Europe as well as the United States and Canada, anyone who logged into their Google account could directly improve the map by fixing incorrect driving directions, adding biking trails, or adding a missing building or road. General map errors in Australia, Austria, Belgium, Denmark, France, Liechtenstein, Netherlands, New Zealand, Norway, South Africa, Switzerland, and the United States could be reported using the Report a Problem link in Google Maps and would be updated by Google. For areas where Google used Tele Atlas data, map errors could be reported using Tele Atlas map insight.

If imagery was missing, outdated, misaligned, or generally incorrect, one could notify Google through their contact request form.

In November 2016, Google announced the discontinuation of Google Map Maker as of March 2017.

Mobile App

Google Maps is available as a mobile app for the Android and iOS mobile operating systems. The Android app was first released in September 2008, though the GPS-localization feature had been in testing on cellphones since 2007. Up until iOS 6, the built-in maps application on the iOS operating system was powered by Google Maps. However, with the announcement of iOS 6 in June 2012, Apple announced that they had created their own Apple Maps mapping service, which officially replaced Google Maps when iOS 6 was released on September 19, 2012. However, at launch, Apple Maps received significant criticism from users due to inaccuracies, errors and bugs. One day later, *The Guardian* reported that Google was preparing its own Google Maps app, which was released on December 12, 2012. Within only two days, the application had been downloaded over ten million times.

Features

The Google Maps apps for iOS and Android have many of the same features, including turn-by-turn navigation, street view, and public transit information. Turn-by-turn navigation was originally announced by Google as a separate beta testing app exclusive to Android 2.0 devices in October 2009. The original standalone iOS version did not support the iPad, but tablet support was added with version 2.0 in July 2013. An update in June 2012 for Android devices added support for offline access to downloaded maps of certain regions, a feature that was eventually released for iOS devices, and made more robust on Android, in May 2014.

At the end of 2015 Google Maps announced its new offline functionality, but with various limitations downloaded area cannot exceed 120,000 square kilometres and require a considerable amount of storage space. In January 2017, Google added a feature exclusively to Android that will, in some U.S. cities, indicate the level of difficulty in finding available parking spots, and on both Android and iOS, the app can, as of an April 2017 update, remember where users parked. In August 2017, Google Maps for Android was updated with new functionality to actively help the user in finding parking lots and garages close to a destination. In December 2017, Google added a new two-wheeler mode to its Android app, designed for users in India, allowing for more accessibility in traffic conditions.

Go Version

Google Maps Go, a version of the app designed for lower-end devices was released in beta in January 2018. By September 2018, the app had over 10 million installations.

Potential Misuse

In 2005 the Australian Nuclear Science and Technology Organization (ANSTO) complained about the potential for terrorists to use the satellite images in planning attacks, with specific reference to the Lucas Heights nuclear reactor; however, the Australian Federal government did not support the organization's concern. At the time of the ANSTO complaint, Google had colored over some areas for security (mostly in the US), such as the rooftop of the White House and several other Washington, D.C., US buildings.

In October 2010, Nicaraguan military commander Edén Pastora stationed Nicaraguan troops on the Isla Calero (in the delta of the San Juan River), justifying his action on the border delineation given by Google Maps. Google has since updated its data which it found to be incorrect.

On January 27, 2014, documents leaked by Edward Snowden revealed that the NSA and the GCHQ intercepted Google Maps queries made on smartphones, and used them to locate the users making these queries. One leaked document, dating to 2008, stated that "it effectively means that anyone using Google Maps on a smartphone is working in support of a GCHQ system."

Google Latitude

Google Latitude was a location-aware feature of Google Maps, developed by Google as a successor to its earlier SMS-based service Dodgeball. Latitude allowed a mobile phone user to allow certain people to view their current location. Via their own Google Account, the user's cell phone location was mapped on Google Maps. The user could control the accuracy and details of what each of the other users can see — an exact location could be allowed, or it could be limited to identifying the city only. For privacy, it could also be turned off by the user, or a location could be manually entered. Users had to explicitly opt into Latitude, and were only able to see the location of those friends who had decided to share their location with them.

On July 10, 2013, Google announced plans to shut down Latitude, and it was discontinued on August 9, 2013. After the feature moved to Google+ in between, Google incorporated Latitude's location sharing feature into Google Maps in March 2017.

Discontinuation

On July 10, 2013 (with less than a month before its closure), Google announced plans to shut down Google Latitude on August 9, 2013. Google then offered location reporting on Google+, but this did not run on all the platforms that Google Maps does (BlackBerry, Windows Mobile, S60, etc.). Later it was fully migrated into Google Maps.

Compatibility

Google Latitude was compatible with most devices running iOS, Android, BlackBerry OS, Windows Mobile, and Symbian S60. Initially Google stated on the Latitude page that it would be available for Java ME phones, but this claim was later removed from the site. On most platforms Latitude could continue to update the user's location in the background when the application was not in use, while on others it only updated the user's location when the application was in use.

The Sony Ericsson W995, C905, C903, C510, Elm and Satio mobile phones supported Google Latitude as part of their built-in Google Maps application. Although this was a Java ME application, it could not be downloaded for use with other mobile phones.

Privacy Concerns

Amid concerns over locational privacy, Google announced that Latitude overwrites a user's previous location with the new location data and does not keep logs of locations provided to the service.

By early 2011, Google Latitude optionally recorded a history of places visited and counts time spent at each place. This information was then used to display statistics such as "Time At Work", "Time Spent At Home" and "Time Spent Out".

GPS Tracking Unit

A GPS tracking unit is a navigation device normally carried by a moving vehicle or person or animal that uses the Global Positioning System (GPS) to track the device's movements and determine its location. The recorded location data can either be stored within the tracking unit or transmitted to an Internet-connected device using the cellular (GPRS or SMS), radio, or satellite modem embedded in the unit. This allows the location to be displayed against a map backdrop either in real time or when analysing the track later, using GPS tracking software. Data tracking software is available for smartphones with GPS capability.

GPS Tracking Unit Architecture

A GPS "track me" essentially contains a GPS module that receives the GPS signal and calculates the coordinates. For data loggers, it contains large memory to store the coordinates. Data pushers additionally contain a GSM/GPRS/CDMA/LTE modem to transmit this information to a central computer either via SMS or GPRS in form of IP packets. Satellite-based GPS tracking units will operate anywhere on the globe using satellite technology such as GlobalStar or Iridium. They do not require a cellular connection.

Types

There are three types of GPS trackers, though most GPS-equipped phones can work in any of these modes depending on the mobile applications installed:

Data Loggers

GPS loggers log the position of the device at regular intervals in its internal memory. GPS loggers may have either a memory card slot, or internal flash memory card and a USB port. Some act as a USB flash drive, which allows downloading of the track log data for further computer analysis. The track list or point of interest list may be in GPX, KML, NMEA or other format.

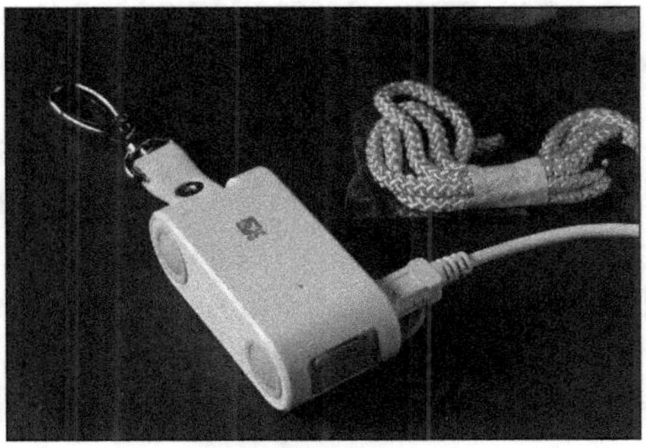

Typical GPS logger.

Most digital cameras save the time a photo was taken. Provided the camera clock is reasonably accurate or used GPS as its time source, this time can be correlated with GPS log data, to provide an accurate location. This can be added to the Exif metadata in the picture file. Cameras with a GPS receiver built in can directly produce such a geotagged photograph.

In some private investigation cases, data loggers are used to keep track of a target vehicle. The private investigator need not follow the target too closely, and always has a backup source of data.

Data Pushers

A data pusher is the most common type of GPS tracking unit, used for asset tracking, personal tracking and vehicle tracking systems.

Also known as a "GPS beacon", this kind of device pushes (i.e. "sends"), at regular intervals, the position of the device as well as other information like speed or altitude to a determined server, that can store and analyze the data instantly.

A GPS navigation device and a mobile phone sit side-by-side in the same box, powered by the same battery. At regular intervals, the phone sends a text message via SMS or GPRS, containing the data from the GPS receiver. Newer GPS-integrated smartphones running GPS tracking software can turn the phone into a data pusher (or logger) device. As of 2009, open source and proprietary applications are available for common Java ME enabled phones, iPhone, Android, Windows Mobile, and Symbian.

Most 21st-century GPS trackers provide data "push" technology, enabling sophisticated GPS tracking in business environments, specifically organizations that employ a mobile workforce, such as a commercial fleet. Typical GPS tracking systems used in commercial fleet management have two core parts: location hardware (or tracking device) and tracking software. This combination is often referred to as an Automatic Vehicle Location system. The tracking device is most often hardwired installed in the vehicle, connected to the CAN-bus, ignition system switch, battery. It allows collection of extra data, which is later transferred to the GPS tracking server. There it is available for viewing, in most cases via a website accessed over the Internet, where fleet activity can be viewed live or historically using digital maps and reports.

GPS tracking systems used in commercial fleets are often configured to transmit location and telemetry input data at a set update rate or when an event (door open/close, auxiliary equipment on/off, geofence border cross) triggers the unit to transmit data. Live GPS tracking used in commercial fleets generally refers to systems that update regularly at one-minute, two-minute or five-minute intervals while the ignition status is on. Some tracking systems combine timed updates with heading change triggered updates.

GPS tracking solutions such as Telematics 2.0, an IoT based telematics technology for the automotive industry, are being used by mainstream commercial auto insurance companies.

The applications of trackers of this kind include:

Personal GPS

- Law enforcement: An arrested suspect out on bail may have to wear a GPS tracker, usually an ankle monitor, as a bail condition. GPS tracking may also be ordered for persons subject to a restraining order.
- Race control: In some sports, such as gliding, participants are required to carry a tracker. In particular, this allows race officials to know if the participants are cheating, taking unexpected shortcuts, and how far apart they are.
- Espionage/surveillance: A tracker on a person or vehicle allows movements to be tracked. This application is used by private investigators.
- Vehicle tracking: Some people use GPS Trackers to monitor activity of their own vehicle, especially in the event of a vehicle being used by a friend or family member.
- GPS personal tracking devices are used in the care of the elderly and vulnerable, and can be used to track small children who may get into danger. Some devices can send text alerts to carers if the wearer moves into an unexpected place. Some devices allow users to call for assistance, and optionally allow designated carers to locate the user's position, typically within five to ten meters. Their use helps promote independent living and social inclusion for the elderly. Devices often incorporate either one-way or two-way voice communication. Some devices also allow the user to call several phone numbers using pre-programmed speed dial buttons. Trials using GPS personal tracking devices for sufferers of early-stage dementia are underway in several countries. Text and voice communication is

usually provided by a connection to mobile telephony, but GPS devices are available that use satellite communications, always available even if out of mobile telephone range.

- Some Internet Web 2.0 pioneers have created their own personal web pages that show their position constantly, and in real time, on a map within their website. These usually use data push from a GPS enabled cell phone or a personal GPS tracker.
- Sports: The movements of a rambler, cyclist, etc., can be tracked. Statistics such as instantaneous and average speed, and distance traveled, are logged.
- Adventure sports: GPS tracking devices such as the SPOT Satellite Messenger are available to allow the position of a person to be tracked. In particular, this allows rescue personnel to locate the carrier. These devices also allow the carrier to send messages and emergency alerts, even when out of cellular telephone range.
- Monitoring employees: GPS-handled tracking devices with a built-in cellphone are used to monitor employees by various companies, especially those engaged in fieldwork.

Asset Tracking

- Solar Powered: The advantage of some solar powered units is that they have much more power over their lifetime than battery powered units. This gives them the advantage of reporting their position and status much more often than battery units which need to conserve energy to extend their life. Some wireless solar-powered units, such as the RailRider can report more than 20,000 times per year and work indefinitely on solar power, eliminating the need to change batteries.
- Battery powered: Asset tracking devices often conserve battery life by reporting in less frequently than vehicle-powered counterparts.
- Animal control: When put on a wild animal (e.g. in a collar), it allows scientists to study the animal's activities and migration patterns. Vaginal implant transmitters mark the location where pregnant females give birth. Animal tracking collars may also be put on domestic animals, to locate them in case they get lost.

Aircraft Trackers

Aircraft can be tracked either by ADS-B (primarily airliners and General Aviation aircraft with ADS-B-out enabled transponder), or by FLARM data packets picked up by a network of ground stations (primarily used by General Aviation aircraft, gliders and UAVs), both of which are data pushers. ADS-B is to be superseded by ADS-C, a data puller.

Data Pullers

GPS data pullers are also known as "GPS transponders". Unlike data pushers that send the position of the devices at regular intervals (push technology), these devices are always on, and can be queried as often as required (pull technology). This technology is not in widespread use, but an example of this kind of device is a computer connected to the Internet and running gpsd.

These can often be used in the case where the location of the tracker will only need to be known occasionally e.g. placed in property that may be stolen, or that does not have a constant source of energy to send data on a regular basis, like freight or containers.

Data Pullers are coming into more common usage in the form of devices containing a GPS receiver and a cell phone which, when sent a special SMS message reply to the message with their location.

Covert GPS Trackers

Covert GPS trackers contain the same electronics as regular GPS trackers but are constructed in such a way as to appear to be an everyday object. One use for covert GPS trackers is for power tool protection, these devices can be concealed within power tool boxes and traced if theft occurs.

An OBD GPS Tracker being inserted into the OBD-II port of a car.

Vehicle OBD GPS Trackers

OBD GPS trackers plug into the onboard diagnostic (OBD) port of a light or medium duty vehicle. Usually, an OBD GPS tracker draws power from the OBD port itself and contains a built-in antenna along with a GPS module to receive the GPS signal. In addition, OBD trackers communicate with the different vehicle subsystems for receiving vehicle diagnostic and fuel consumption related data. A cellular OBD GPS tracker directly communicates with the cell tower to send the location and other vehicle performance data to the server over a cellular wireless network. Users can view the information using standalone software or a web browser from a desktop/laptop computer or using smartphone apps.

Uses in Marketing

In August 2010, Brazilian company Unilever ran an unusual promotion where GPS trackers were placed in boxes of Omo laundry detergent. Teams would then track consumers who purchased the boxes of detergent to their homes where they would be awarded a prize for their purchase. The company also launched a website to show the approximate location of the winners' homes.

GPS Logger for Android

A battery efficient GPS logger application for Android that works on Android 1.5 onwards. A guide for OSM logging is available. GPS logger supports uploading to OSM, at regular intervals, when a track ends, and/or manually.

GPS Logger uses the GPS capabilities of an Android phone to log coordinates to a GPX file at custom time/distance intervals. This can be particularly useful for geotagging photos after a day out or share travel routes with someone. The purpose of this application is to be battery efficient to save battery power when abroad and last as long as possible.

A useful feature is the "passive" location source, which listens for locations acquired by other apps instead of asking for locations actively. This can be helpful when another app that also uses location is running. Such apps may include street-view collection for Mapillary/OpenStreetView records, wardriving with Mozilla Stumbler/WiGLE, or simply an AR game based on GPS.

5
Geolocation

Geolocation is the process of estimating and generating the geographical coordinates of the real-world geographic location of an object such as a radar source, mobile phone and Internet connected computer terminal. Indoor positioning system, Wi-Fi positioning system, geographic system, etc. are some of the systems used in it. This chapter discusses the subject of geolocation in detail.

Geolocation is the process of finding, determining and providing the exact location of a computer, networking device or equipment. It enables device location based on geographical coordinates and measurements.

Geolocation commonly uses Global Positioning System (GPS) and other related technologies to assess and specify geographical locations.

Geolocation provides the location of a device but is generally used in a variety of applications to help locate human users. Geolocation works through a pre-built GPS in a device that propagates the device's longitudinal and latitudinal coordinates. The coordinates are identified on a map to provide a complete address that usually includes a country, city, town/colony, building name and street address.

Besides GPS, geolocation also may be identified through an Internet Protocol (IP) address, media access control (MAC) address, radio frequency (RF) systems, Exchangeable Image File Format (EXIF) data and other wireless positioning systems.

Radio Navigation

Radio navigation or radionavigation is the application of radiofrequencies to determine a position of an object on the Earth. Like radiolocation, it is a type of radiodetermination.

The basic principles are measurements from/to electric beacons, especially:

- Angular directions, e.g. by bearing, radio phases or interferometry.

- Distances, e.g. ranging by measurement of time of flight between one transmitter and multiple receivers or vice versa.
- Distance *differences* by measurement of times of arrival of signals from one transmitter to multiple receivers or vice versa.
- Partly also velocity, e.g. by means of radio Doppler shift.

Combinations of these measurement principles also are important—e.g., many radars measure range and azimuth of a target.

Bearing-measurement Systems

These systems used some form of directional radio antenna to determine the location of a broadcast station on the ground. Conventional navigation techniques are then used to take a radio fix. These were introduced prior to World War I, and remain in use today.

Radio Direction Finding

Amelia Earhart's Lockheed Electrahad a prominent RDF loop on the cockpit roof.

The first system of radio navigation was the *Radio Direction Finder*, or RDF. By tuning in a radio stationand then using a directional antenna, one could determine the direction to the broadcasting antenna. A second measurement using another station was then taken. Using triangulation, the two directions can be plotted on a map where their intersection reveals the location of the navigator. Commercial AM radio stations can be used for this task due to their long range and high power, but strings of low-power radio beacons were also set up specifically for this task, especially near airports and harbours.

Early RDF systems normally used a loop antenna, a small loop of metal wire that is mounted so it can be rotated around a vertical axis. At most angles the loop has a fairly flat reception pattern, but when it is aligned perpendicular to the station the signal received on one side of the loop cancels the signal in the other, producing a sharp drop in reception known as the "null". By rotating the loop and looking for the angle of the null, the relative bearing of the station can be determined. Loop antennas can be seen on most pre-1950s aircraft and ships.

Reverse RDF

The main problem with RDF is that it required a special antenna on the vehicle, which may not be easy to mount on smaller vehicles or single-crew aircraft. A smaller problem is that the accuracy of the system is based to a degree on the size of the antenna, but larger antennas would likewise make the installation more difficult.

The Orfordness Beacon as it appears today.

During the era between World War I and World War II, a number of systems were introduced that placed the rotating antenna on the ground. As the antenna rotated through a fixed position, typically due north, the antenna was keyed with the morse code signal of the station's identification letters so the receiver could ensure they were listening to the right station. Then they waited for the signal to either peak or disappear as the antenna briefly pointed in their direction. By timing the delay between the morse signal and the peak/null, then dividing by the known rotational rate of the station, the bearing of the station could be calculated.

The first such system was the German Telefunken Kompass Sender, which began operations in 1907 and was used operationally by the Zeppelin fleet until 1918. An improved version was introduced by the UK as the Orfordness Beacon in 1929 and used until the mid-1930s. A number of improved versions followed, replacing the mechanical motion of the antennas with phasing techniques that produced the same output pattern with no moving parts. One of the longest lasting examples was Sonne, which went into operation just before World War II and was used operationally under the name Consol until 1991. The modern VOR system is based on the same principles.

ADF and NDB

A great advance in the RDF technique was introduced in the form of phase comparisons of a signal as measured on two or more small antennas, or a single highly directional solenoid. These receivers were smaller, more accurate, and simpler to operate. Combined with the introduction of the transistor and integrated circuit, RDF systems were so reduced in size and complexity that they once again became quite common during the 1960s, and were known by the new name, automatic direction finder, or ADF.

This also led to a revival in the operation of simple radio beacons for use with these RDF systems, now referred to as *non-directional beacons* (NDB). As the LF/MF signals used by NDBs can follow the curvature of earth, NDB has a much greater range than VOR which travels only in *line of sight*. NDB can be categorized as *long range* or *short range* depending on their power. The frequency band allotted to non-directional beacons is 190–1750 kHz, but the same system can be used with any common AM-band commercial station.

VOR

VOR transmitter station.

VHF omnidirectional range, or VOR, is an implementation of the reverse-RDF system, but one that is more accurate and able to be completely automated.

The VOR station transmits two audio signals on a VHF carrier one is at 1020 Hz which is Morse code to identify the station, the other is a continuous 9960 Hz audio signal frequency modulated at 30 Hz, with the 0-degree referenced to magnetic north. This signal is rotated mechanically or electrically at 30 Hz, which appears as a 30 Hz AM signal added to the previous two signals, the phasing of which is dependent on the position of the aircraft relative to the VOR station.

In the VOR signal is a single RF carrier that is demodulated into a composite audio signal composed of a 9960 Hz reference signal frequency modulated at 30 Hz, a 30 Hz AM reference signal, and a 1020 Hz 'marker' signal for station identification. Conversion from this audio signal into a usable navigation aid is done by a navigation converter, which takes the reference signal and compares the phasing with the variable signal. Station identification is known by listening to the audio directly, as the 9960 Hz and 30 Hz signals are filtered out of the aircraft internal communication system, leaving only the 1020 Hz Morse-code station identification. The phase difference in degrees is transmitted to flight station navigational displays for use by the flight crew.

The system is often able to be utilized with a compatible glideslope and marker beacon receiver, making the aircraft ILS-capable (Instrument Landing System). Once the aircraft's approach is accurate (the aircraft is in the "right place"), the VOR receiver will be used on a different frequency to determine if the aircraft is pointed in the "right direction". Military aircraft will usually employ two VOR receiver systems, one in VOR-only mode to determine "right place" and another in ILS mode in conjunction with a glideslope receiver to determine "right direction". The combination of both allows for a precision approach in foul weather.

Beam Systems

Beam systems broadcast narrow signals in the sky, and navigation is accomplished by keeping the aircraft centred in the beam. A number of stations are used to create an airway, with the navigator tuning in different stations along the direction of travel. These systems were common in the era when electronics were large and expensive, as they placed minimum requirements on the receivers they were simply voice radio sets tuned to the selected frequencies. However, they did not provide navigation outside of the beams, and were thus less flexible in use. The rapid miniaturization of electronics during and after World War II made systems like VOR practical, and most beam systems rapidly disappeared.

Lorenz

In the post-World War I era, the Lorenz company of Germany developed a means of projecting two narrow radio signals with a slight overlap in the center. By broadcasting different audio signals in the two beams, the receiver could position themselves very accurately down the centreline by listening to the signal in their headphones. The system was accurate to less than a degree in some forms.

Originally known as "Ultrakurzwellen-Landefunkfeuer" (LFF), or simply "Leitstrahl" (guiding beam), little money was available to develop a network of stations. Deployment was instead led by the US, where it formed the basis of a wide-area navigation system through the 1930s and 40s. Development was restarted in Germany in the 1930s as a short-range system deployed at airports as a blind landing aid. Although there was some interest in deploying a medium-range system like the US LFF, deployment had not yet started when the beam system was combined with the Orfordness timing concepts to produce the highly accurate Sonne system. In all of these roles, the system was generically known simply as a "Lorenz beam".

In the immediate pre-World War II era the same concept was also developed as a blind-bombing system. This used very large antennas to provide the required accuracy at long distances (over England), and very powerful transmitters. Two such beams were used, crossing over the target to triangulate it. Bombers would enter one of the beams and use it for guidance until they heard the second one in a second radio receiver, using that signal to time the dropping of their bombs. The system was highly accurate, and the 'Battle of the Beams' broke out when United Kingdom intelligence servicesattempted, and then succeeded, in rendering the system useless through electronic warfare. Sonne, however, proved just as useful to the UK as Germany, and was left to operate unhindered throughout the war.

Low Frequency Radio Range

LFR ground station.

The low-frequency radio range (LFR, also other names) was the main navigation system used by aircraft for instrument flying in the 1930s and 1940s in the U.S. and other countries, until the advent of the VOR in the late 1940s. It was used for both en route navigation as well as instrument approaches.

The ground stations consisted of a set of four antennas that projected Lorenz beams in four cardinal directions. One of the beams was "keyed" with the morse code signal "A", dit-dah, with the second beam "N", dah-dit. Flying down the centreline produced a steady tone. The beams were pointed to the next station to produce a set of airways, allowing an aircraft to travel from airport to airport by following a selected set of stations. Effective course accuracy was about three degrees, which near the station provided sufficient safety margins for instrument approaches down to low minimums. At its peak deployment, there were nearly 400 LFR stations in the US.

Glide Path and the Localizer of ILS

The remaining widely used beam systems are glide path and the localizer of the *instrument landing system* (ILS). ILS uses a *localizer* to provide horizontal position and *glide path* to provide vertical positioning. ILS can provide enough accuracy and redundancy to allow automated landings.

Transponder Systems

Positions can be determined with any two measures of angle or distance. The introduction of radar in the 1930s provided a way to directly determine the distance to an object even at long distances. Navigation systems based on these concepts soon appeared, and remained in widespread use until recently. Today they are used primarily for aviation, although GPS has largely supplanted this role.

Radar and Transponders

Early radar systems, like the UK's Chain Home, consisted of large transmitters and separate receivers. The transmitter periodically sends out a short pulse of a powerful radio signal, which is sent into space through broadcast antennas. When the signal reflects off a target, some of that signal is reflected back in the direction of the station, where it is received. The received signal is a tiny fraction of the broadcast power, and has to be powerfully amplified in order to be used.

The same signals are also sent over local electrical wiring to the operator's station, which is equipped with an oscilloscope. Electronics attached to the oscilloscope provides a signal that increases in voltage over a short period of time, a few microseconds. When sent to the X input of the oscilloscope, this causes a horizontal line to be displayed on the scope. This "sweep" is triggered by a signal tapped off the broadcaster, so the sweep begins when the pulse is sent. Amplified signals from the receiver are then sent to the Y input, where any received reflection causes the beam to move upward on the display. This causes a series of "blips" to appear along the horizontal axis, indicating reflected signals. By measuring the distance from the start of the sweep to the blip, which corresponds to the time between broadcast and reception, the distance to the object can be determined.

Soon after the introduction of radar, the radio transponder appeared. Transponders are a combination of receiver and transmitter whose operation is automated upon reception of a particular signal, normally a pulse on a particular frequency, the transponder sends out a pulse in response, typically delayed by some very short time. Transponders were initially used as the basis for early IFF systems; aircraft with the proper transponder would appear on the display as part of the normal radar operation, but then the signal from the transponder would cause a second blip to appear a short time later. Single blips were enemies, double blips friendly.

Transponder-based distance-distance navigation systems have a significant advantage in terms of positional accuracy. Any radio signal spreads out over distance, forming the fan-like beams of the Lorenz signal, for instance. As the distance between the broadcaster and receiver grows, the area covered by the fan increases, decreasing the accuracy of location within it. In comparison, transponder-based systems measure the timing between two signals, and the accuracy of that measure is largely a function of the equipment and nothing else. This allows these systems to remain accurate over very long range.

The latest transponder systems (mode S) can also provide position information, possibly derived from GNSS, allowing for even more precise positioning of targets.

Bombing Systems

The first distance-based navigation system was the German Y-Gerät blind-bombing system. This used a Lorenz beam for horizontal positioning, and a transponder for ranging. A ground-based system periodically sent out pulses which the airborne transponder returned. By measuring the total round-trip time on a radar's oscilloscope, the aircraft's range could be accurately determined even at very long ranges. An operator then relayed this information to the bomber crew over voice channels, and indicated when to drop the bombs.

The British introduced similar systems, notably the Oboe system. This used two stations in England that operated on different frequencies and allowed the aircraft to be triangulated in space. To ease pilot workload only one of these was used for navigation prior to the mission a circle was drawn over the target from one of the stations, and the aircraft was directed to fly along this circle on instructions from the ground operator. The second station was used, as in Y-Gerät, to time the bomb drop. Unlike Y-Gerät, Oboe was deliberately built to offer very high accuracy, as good as 35 m, much better than even the best optical bombsights.

One problem with Oboe was that it allowed only one aircraft to be guided at a time. This was addressed in the later Gee-H system by placing the transponder on the ground and broadcaster in the aircraft. The signals were then examined on existing Gee display units in the aircraft. Gee-H did not offer the accuracy of Oboe, but could be used by as many as 90 aircraft at once. This basic concept has formed the basis of most distance measuring navigation systems to this day.

Beacons

The key to the transponder concept is that it can be used with existing radar systems. The ASV radar introduced by RAF Coastal Command was designed to track down submarines and ships by displaying the signal from two antennas side by side and allowing the operator to compare their relative strength. Adding a ground-based transponder immediately turned the same display into a system able to guide the aircraft towards a transponder, or "beacon" in this role, with high accuracy.

The British put this concept to use in their Rebecca/Eureka system, where battery-powered "Eureka" transponders were triggered by airborne "Rebecca" radios and then displayed on ASV Mk. II radar sets. Eureka's were provided to French resistance fighters, who used them to call in supply drops with high accuracy. The US quickly adopted the system for paratroop operations, dropping the Eureka with pathfinder forces or partisans, and then homing in on those signals to mark the drop zones.

The beacon system was widely used in the post-war era for blind bombing systems. Of particular note were systems used by the US Marines that allowed the signal to be delayed in such a way to offset the drop point. These systems allowed the troops at the front line to direct the aircraft to points in front of them, directing fire on the enemy. Beacons were widely used for temporary or mobile navigation as well, as the transponder systems were generally small and low-powered, able to be man portable or mounted on a Jeep.

DME

In the post-war era, a general navigation system using transponder-based systems was deployed as the distance measuring equipment (DME) system.

DME was identical to Gee-H in concept, but used new electronics to automatically measure the time delay and display it as a number, rather than having the operator time the signals manually on an oscilloscope. This led to the possibility that DME interrogation pulses from different aircraft might be confused, but this was solved by having each aircraft send out a different series of pulses which the ground-based transponder repeated back.

DME is almost always used in conjunction with VOR, and is normally co-located at a VOR station. This combination allows a single VOR/DME station to provide both angle and distance, and thereby provide a single-station fix. DME is also used as the distance-measuring basis for the military TACAN system, and their DME signals can be used by civilian receivers.

Hyperbolic Systems

Hyperbolic navigation systems are a modified form of transponder systems which eliminate the need for an airborne transponder. The name refers to the fact that they do not produce a single distance or angle, but instead indicate a location along any number of hyperbolic lines in space. Two such measurements produce a fix. As these systems are almost always used with a specific navigational chart with the hyperbolic lines plotted on it, they generally reveal the receiver's location directly, eliminating the need for manual triangulation. As these charts were digitized, they became the first true location-indication navigational systems, outputting the location of the receiver as latitude and longitude. Hyperbolic systems were introduced during World War II and remained the main long-range advanced navigation systems until GPS replaced them in the 1990s.

Gee

The first hyperbolic system to be developed was the British Gee system, developed during World War II. Gee used a series of transmitters sending out precisely timed signals, with the signals leaving the stations at fixed delays. An aircraft using Gee, RAF Bomber Command's heavy bombers, examined the time of arrival on an oscilloscope at the navigator's station. If the signal from two stations arrived at the same time, the aircraft must be an equal distance from both transmitters, allowing the navigator to determine a line of position on his chart of all the positions at that distance from both stations. More typically, the signal from one station would be received earlier than the other. The *difference* in timing between the two signals would reveal them to be along a curve of possible locations. By making similar measurements with other stations, additional lines of position can be produced, leading to a fix. Gee was accurate to about 165 yards (150 m) at short ranges, and up to a mile (1.6 km) at longer ranges over Germany. Gee remained in use long after World War II, and equipped RAF aircraft as late as the 1960s (approx. frequency was by then 68 MHz).

LORAN

With Gee entering operation in 1942, similar US efforts were seen to be superfluous. They turned their development efforts towards a much longer-ranged system based on the same principles, using much lower frequencies that allowed coverage across the Atlantic Ocean. The result was LORAN, for "LOng-range Aid to Navigation". The downside to the long-wavelength approach was that accuracy was greatly reduced compared to the high-frequency Gee. LORAN was widely used during convoy operations in the late war period.

Decca

Another British system from the same era was Decca Navigator. This differed from Gee primarily in that the signals were not pulses delayed in time, but continuous signals delayed in phase. By comparing the phase of the two signals, the time difference information as Gee was returned. However, this was far easier to display; the system could output the phase angle to a pointer on a dial removing any need for visual interpretation. As the circuitry for driving this display was quite small, Decca systems normally used three such displays, allowing quick and accurate reading of multiple fixes. Decca found its greatest use post-war on ships, and remained in use into the 1990s.

LORAN-C

Almost immediately after the introduction of LORAN, in 1952 work started on a greatly improved version. LORAN-C (the original retroactively became LORAN-A) combined the techniques of pulse timing in Gee with the phase comparison of Decca.

The resulting system (operating in the low frequency (LF) radio spectrum from 90 to 110 kHz) that was both long-ranged (for 60 kW stations, up to 3400 miles) and accurate. To do this, LORAN-C sent a pulsed signal, but modulated the pulses with an AM signal within it. Gross positioning was determined using the same methods as Gee, locating the receiver within a wide area. Finer accuracy was then provided by measuring the phase difference of the signals, overlaying that second measure on the first. By 1962, high-power LORAN-C was in place in at least 15 countries.

LORAN-C was fairly complex to use, requiring a room of equipment to pull out the different signals. However, with the introduction of integrated circuits, this was quickly reduced further and further. By the late 1970s, LORAN-C units were the size of a stereo amplifier and were commonly found on almost all commercial ships as well as some larger aircraft. By the 1980s, this had been further reduced to the size of a conventional radio, and it became common even on pleasure boats and personal aircraft. It was the most popular navigation system in use through the 1980s and 90s, and its popularity led to many older systems being shut down, like Gee and Decca. However, like the beam systems before it, civilian use of LORAN-C was short-lived when GPS technology drove it from the market.

Other Hyperbolic Systems

Similar hyperbolic systems included the US global-wide VLF/Omega Navigation System, and the similar Alpha deployed by the USSR. These systems determined pulse timing not by comparison of two signals, but by comparison of a single signal with a local atomic clock. The expensive-to-maintain Omega system was shut down in 1997 as the US military migrated to using GPS. Alpha is still in use.

Satellite Navigation

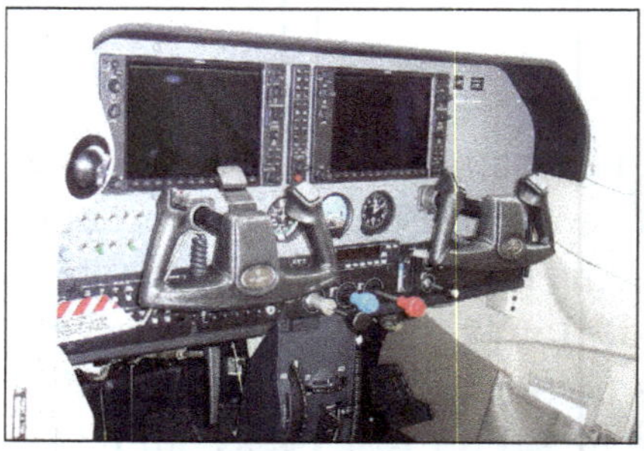

Cessna 182 with GPS-based "glass cockpit" avionics.

Since the 1960s, navigation has increasingly moved to satellite navigation systems. These are essentially hyperbolic systems whose transmitters are in orbits. That the satellites move with respect to the receiver requires that the calculation of the positions of the satellites must be taken into account, which can only be handled effectively with a computer.

Satellite navigation systems send several signals that are used to decode the satellite's position, distance between the user satellite, and the user's precise time. One signal encodes the satellite's ephemeris data, which is used to accurately calculate the satellite's location at any time. Space weather and other effects causes the orbit to change over time so the ephemeris has to be updated periodically. Other signals send out the time as measured by the satellite's onboard atomic clock. By measuring signal times of arrival (TOAs) from at least four satellites, the user's receiver can re-build an accurate clock signal of its own and allows hyperbolic navigation to be carried out.

Satellite navigation systems offer better accuracy than any land-based system, are available at almost all locations on the Earth, can be implemented (receiver-side) a modest cost and complexity, with modern electronics, and require only a few dozen satellites to provide worldwide coverage. As a result of these advantages, satellite navigation has led to almost all previous systems falling from use. LORAN, Omega, Decca, Consol and many other systems disappeared during the 1990s and 2000s. The only other systems still in use are aviation aids, which are also being turned off for long-range navigation while new differential GPS systems are being deployed to provide the local accuracy needed for blind landings.

Radiolocation

Radiolocating is the process of finding the location of something through the use of radio waves. It generally refers to passive uses, particularly radar—as well as detecting buried cables, water mains, and other public utilities. It is similar to radionavigation, but radiolocation usually refers to passively finding a distant object rather than actively one's own position. Both are types of radiodetermination. Radiolocation is also used in real-time locating systems (RTLS) for tracking valuable assets.

Basic Principles

An object can be located by measuring the characteristics of received radio waves. The radio waves may be transmitted by the object to be located, or they may be backscattered waves (as in radar or passive RFID). A finder uses radiolocation when it uses radio waves rather than ultrasound.

One technique measures a distance by using the difference in the power of the received signal strength (RSSI) as compared to the originating signal strength. Another technique uses the time of arrival (TOA), when the time of transmission and speed of propagation are known. Combining TOA data from several receivers at different known locations (time difference of arrival, TDOA) can provide an estimate of position even in the absence of knowledge of the time of transmission. The angle of arrival (AOA) at a receiving station can be determined by the use of a directional antenna, or by differential time of arrival at an array of antennas with known location. Estimates are improved when the transmission characteristics of the medium is factored into the calculations. For RSSI this means electromagnetic permeability; for TOA it may mean non-line-of-sight receptions.

Use of RSSI to locate a transmitter from a single receiver requires that both the transmitted (or backscattered) power from the object to be located is known, and that the propagation characteristics of the intervening region are known. In empty space, signal strength decreases as the inverse square of the distance for distances large compared to a wavelength and compared to the object to be located, but in most real environments, a number of impairments can occur: absorption, refraction, shadowing, and reflection. Absorption is negligible for radio propagation in air at frequencies less than about 10 GHz, but becomes important at multi-GHz frequencies where rotational molecular states can be excited. Refraction is important at long ranges (tens to hundreds of kilometers) due to gradients in moisture content and temperature in the atmosphere. In urban, mountainous, or indoor environments, obstruction by intervening obstacles and reflection from nearby surfaces are very common, and contribute to multipath distortion: that is, reflected and delayed replicates of the transmitted signal are combined at the receiver. Signals from different paths can add constructively or destructively: such variations in amplitude are known as fading. The dependence of signal strength on position of transmitter and receiver becomes complex and often non-monotonic, making single-receiver estimates of position inaccurate and unreliable. Multilateration using many receivers is often combined with calibration measurements ("fingerprinting") to improve accuracy.

TOA and AOA measurements are also subject to multipath errors, particularly when the direct path from the transmitter to receiver is blocked by an obstacle. Time of arrival measurements are also most accurate when the signal has distinct time-dependent features on the scale of interest—for example, when it is composed of short pulses of known duration—but Fourier transform theory shows that in order to change amplitude or phase on a short time scale, a signal must use a broad bandwidth. For example, to create a pulse of about 1 ns duration, roughly sufficient to identify location to within 0.3 m (1 foot), a bandwidth of roughly 1 GHz is required. In many regions of the radio spectrum, emission over such a broad bandwidth is not allowed by the relevant regulatory authorities, in order to avoid interference with other narrowband users of the spectrum. In the United States, unlicensed transmission is allowed in several bands, such as the 902-928 MHz and 2.4-2.483 GHz Industrial, Scientific, and Medical ISM bands, but high-power transmission cannot extend outside of these bands. However, several jurisdictions now allow ultrawideband transmission over GHz or multi-GHz bandwidths, with constraints on transmitted power to minimize interference with other spectrum users. UWB pulses can be very narrow in time, and often provide accurate estimates of TOA in urban or indoor environments.

Radiolocation is employed in a wide variety of industrial and military activities. Radar systems often use a combination of TOA and AOA to determine a backscattering object's position using a single receiver. In Doppler radar, the Doppler shift is also taken into account, determining velocity rather than location (though it helps determine future location). Real Time Location Systems RTLS using calibrated RTLS, and TDOA, are commercially available. The widely used Global Positioning System (GPS) is based on TOA of signals from satellites at known positions.

Mobile Phones

Radiolocation is also used in cellular telephony via base stations. Most often, this is done through trilateration between radio towers. The location of the Caller or handset can be determined several ways:

- Angle of arrival (AOA) requires at least two towers, locating the caller at the point where the lines along the angles from each tower intersect.
- Time difference of arrival (TDOA) resp. time of arrival (TOA) works using multilateration, except that it is the networks that determine the time difference and therefore distance from each tower (as with seismometers).
- Location signature uses "fingerprinting" to store and recall patterns (such

as multipath) which mobile phone signals are known to exhibit at different locations in each cell.

The first two depend on a line-of-sight, which can be difficult or impossible in mountainous terrain or around skyscrapers. Location signatures actually work *better* in these conditions however. TDMA and GSM networks such as Cingular and T-Mobile use TDOA.

CDMA networks such as Verizon Wireless and Sprint PCS tend to use handset-based radiolocation technologies, which are technically more similar to radionavigation. GPS is one of those technologies.

Composite solutions, needing both the handset and the network include:

- Assisted GPS (wireless or TV) allows use of GPS even indoors.
- Advanced Forward Link Trilateration (A-FLT).
- Timing Advance/Network Measurement Report (TA/NMR).
- Enhanced Observed Time Difference (E-OTD).

Initially, the purpose of any of these in mobile phones is so that the public safety answering point (PSAP) which answers calls to an emergency telephone number can know where the caller is and exactly where to send emergency services. This ability is known within the NANP (North America) as wireless enhanced 911. Mobile phone users may have the option to permit the location information gathered to be sent to other phone numbers or data networks, so that it can help people who are simply lost or want other location-based services. By default, this selection is usually turned off, to protect privacy.

Real-time Locating System

Real-time locating systems (RTLS) are used to automatically identify and track the location of objects or people in real time, usually within a building or other contained area. Wireless RTLS tags are attached to objects or worn by people, and in most RTLS, fixed reference points receive wireless signals from tags to determine their location. Examples of real-time locating systems include tracking automobiles through an assembly line, locating pallets of merchandise in a warehouse, or finding medical equipment in a hospital.

The physical layer of RTLS technology is usually some form of radio frequency (RF) communication, but some systems use optical (usually infrared) or acoustic (usually ultrasound) technology instead of or in addition to RF. Tags and fixed reference points can be transmitters, receivers, or both, resulting in numerous possible technology combinations.

RTLS are a form of local positioning system, and do not usually refer to GPSor to mobile phone tracking. Location information usually does not include speed, direction, or spatial orientation.

RTLS are generally used in indoor and/or confined areas, such as buildings, and do not provide global coverage like GPS. RTLS tags are affixed to mobile items to be tracked or managed. RTLS reference points, which can be either transmitters or receivers, are spaced throughout a building (or similar area of interest) to provide the desired tag coverage. In most cases, the more RTLS reference points that are installed, the better the location accuracy, until the technology limitations are reached.

A number of disparate system designs are all referred to as "real-time locating systems", but there are two primary system design elements.

Locating at Choke Points

The simplest form of choke point locating is where short range ID signals from a moving tag are received by a single fixed reader in a sensory network, thus indicating the location coincidence of reader and tag. Alternately, a choke point identifier can be received by the moving tag, and then relayed, usually via a second wireless channel, to a location processor. Accuracy is usually defined by the sphere spanned with the reach of the choke point transmitter or receiver. The use of directional antennas, or technologies such as infrared or ultrasound that are blocked by room partitions, can support choke points of various geometries.

Locating in Relative Coordinates

ID signals from a tag is received by a multiplicity of readers in a sensory network, and a position is estimated using one or more locating algorithms, such as trilateration, multilateration, or triangulation. Equivalently, ID signals from several RTLS reference points can be received by a tag, and relayed back to a location processor. Localization with multiple reference points requires that distances between reference points in the sensory network be known in order to precisely locate a tag, and the determination of distances is called ranging.

Another way to calculate relative location is if mobile tags communicate directly with each other, then relay this information to a location processor.

Location Accuracy

RF trilateration uses estimated ranges from multiple receivers to estimate the location of a tag. RF triangulation uses the angles at which the RF signals arrive at multiple receivers to estimate the location of a tag. Many obstructions, such as walls or furniture, can distort the estimated range and angle readings leading to varied qualities of location estimate. Estimation-based locating is often measured in accuracy for a given distance, such as 90% accurate for 10 meter range.

Systems that use locating technologies that do not go through walls, such as infrared or ultrasound, tend to be more accurate in an indoor environment because only tags and receivers that have line of sight (or near line of sight) can communicate.

Applications

RTLS can be used numerous logistical or operational areas such as:

- Locate and manage assets within a facility, such as finding a misplaced tool cart in a warehouse or medical equipment.
- Notification of new locations, such as an alert if a tool cart improperly has left the facility.
- To combine identity of multiple items placed in a single location, such as on a pallet.
- To locate customers, for example in a restaurant, for delivery of food or service.
- To maintain proper staffing levels of operational areas, such as ensuring guards are in the proper locations in a correctional facility.
- To quickly and automatically account for all staff after or during an emergency evacuation.
    - Toronto General Hospital is looking at RTLS to reduce quarantine times after an infectious disease outbreak. After a recent SARSoutbreak, 1% of all staff were quarantined, and more accurate data regarding who had been exposed to the virus could have reduced the need for quarantines.
- To automatically track and time stamp the progress of people or assets through a process, such as following a patient's emergency room wait time, time spent in the operating room, and total time until discharge. Such a system can be used for process improvement.
- Clinical-grade locating to support acute care capacity management.

Privacy Concerns

RTLS may be seen as a threat to privacy when used to determine the location of people. The newly declared human right of informational self-determination gives the right to prevent one's identity and personal data from being disclosed to others, and also covers disclosure of locality, though this does not generally apply to the workplace.

Several prominent labor unions have come out against the use of RTLS systems to track workers calling them "the beginning of Big Brother" and "an invasion of privacy".

Types of Technologies Used

There is a wide variety of systems concepts and designs to provide real-time locating:

- Active radio frequency identification (Active RFID),
- Active radio frequency identification infrared hybrid (Active RFID-IR),
- Infrared (IR),
- Optical locating,
- Low-frequency signpost identification,
- Semi-active radio frequency identification (semi-active RFID),
- Passive RFID RTLS locating via Steerable Phased Array Antennae,
- Radio beacon,
- Ultrasound Identification (US-ID),
- Ultrasonic ranging (US-RTLS),
- Ultra-wideband (UWB),
- Wide-over-narrow band,
- Wireless Local Area Network (WLAN, Wi-Fi),
- Bluetooth,
- Clustering in noisy ambience,
- Bivalent systems.

A general model for selection of the best solution for a locating problem has been constructed at the Radboud University of Nijmegen. Many of these references do not comply with the definitions given in international standardization with ISO/IEC 19762-5 and ISO/IEC 24730-1. However, some aspects of real-time performance are served and aspects of locating are addressed in context of absolute coordinates.

Ranging and Angulating

Depending on the physical technology used, at least one and often some combination of ranging and/or angulating methods are used to determine location:

- Angle of Arrival (AoA),

- Angle of Departure (AoD) (e.g. Bluetooth Direction Finding, features a mobile-centric RTLS architecture),
- Line-of-sight (LoS),
- Time of arrival (ToA),
- Multilateration (Time difference of arrival) (TDoA),
- Time-of-flight (ToF),
- Two-way ranging (TWR) according to Nanotron's patents,
- Symmetrical Double Sided – Two Way Ranging (SDS-TWR),
- Near-field electromagnetic ranging (NFER).

Errors and Accuracy

Real-time locating is affected by a variety of errors. Many of the major reasons relate to the physics of the locating system, and may not be reduced by improving the technical equipment.

None or no Direct Response

Many RTLS systems require direct and clear line of sight visibility. For those systems, where there is no visibility from mobile tags to fixed nodes there will be no result or a non-valid result from locating engine. This applies to satellite locating as well as other RTLS systems such as angle of arrival and time of arrival. Fingerprinting is a way to overcome the visibility issue: If the locations in the tracking area contain distinct measurement fingerprints, line of sight is not necessarily needed. For example, if each location contains a unique combination of signal strength readings from transmitters, the location system will function properly. This is true, for example, with some Wi-Fi based RTLS solutions. However, having distinct signal strength fingerprints in each location typically requires a fairly high saturation of transmitters.

False Location

The measured location may appear entirely faulty. This is a generally result of simple operational models to compensate for the plurality of error sources. It proves impossible to serve proper location after ignoring the errors.

Locating Backlog

Real time is no registered branding and has no inherent quality. A variety of offers sails under this term. As motion causes location changes, inevitably the latency time to compute a new location may be dominant with regard to motion. Either an RTLS system that requires waiting for new results is not worth the money or the operational concept that asks for faster location updates does not comply with the chosen systems approach.

Temporary Location Error

Location will never be reported *exactly*, as the term *real-time* and the term *precision* directly contradict in aspects of measurement theory as well as the term *precision* and the term *cost* contradict in aspects of economy. That is no exclusion of precision, but the limitations with higher speed are inevitable.

Steady Location Error

Recognizing a reported location steadily apart from physical presence generally indicates the problem of insufficient over-determination and missing of visibility along at least one link from resident anchors to mobile transponders. Such effect is caused also by insufficient concepts to compensate for calibration needs.

Location Jitter

Noise from various sources has an erratic influence on stability of results. The aim to provide a steady appearance increases the latency contradicting to real time requirements.

Location Jump

As objects containing mass have limitations to jump, such effects are mostly beyond physical reality. Jumps of reported location not visible with the object itself generally indicate improper modeling with the location engine. Such effect is caused by changing dominance of various secondary responses.

Location Creep

Location of residing objects gets reported moving, as soon as the measures taken are biased by secondary path reflections with increasing weight over time. Such effect is caused by simple averaging and the effect indicates insufficient discrimination of first echoes.

Automatic Vehicle Location

Automatic vehicle location is a means for determining the geographic location of a vehicle and transmitting this information to a point where it can be used.

Most commonly, the location is determined using GPS and the transmission mechanism is a satellite, terrestrial radio or cellular connection from the vehicle to a radio receiver, satellite or nearby cell tower.

The tracking data is then transmitted using any one of a variety of telemetry systems. GSM and EVDO are the most common technologies used for telemetry because of the low data rate needed for AVL. The low bandwidth requirements also allow for satellite technology to receive telemetry data at a moderately higher cost, but across a global coverage area and into very remote locations not well-covered by terrestrial radio or public carriers.

Automatic Vehicle Location is a Powerful Tool for Managing Fleets

Vehicles such as service vehicles, emergency vehicles, and construction equipment and public transport vehicles (buses and trains) are all using automatic vehicle location (AVL) systems or GPS tracking systems. It also is used to track mobile remote assets such as construction equipment, trailers, and portable power generators.

Types of AVL Systems

Direction Finding

Amateur radio and some cellular or PCS wireless systems use direction finding or triangulation of transmitter signals radiated by the mobile. This is sometimes called radio direction finding, or RDF. The simplest forms of these systems calculate the bearing from two fixed sites to the mobile. This creates a triangle with endpoints at the two fixed points and the mobile. Trigonometry tells you roughly where the mobile transmitter is located. In wireless telephone systems, the phones transmit continually when off-hook, making continual tracking.

Signpost Systems

To track and locate vehicles along fixed routes, a technology called signpost transmitters was employed. This is used on transit routes and rail lines where the vehicles to be tracked continually operated on the same linear route. A transponder or RFID chip along the vehicle route would be polled as the train or bus traverses its route. As each transponder was passed, the moving vehicle would query and receive an ark, or handshake, from the signpost transmitter. A transmitter on the mobile would report passing the signpost to a system controller. These systems are an alternative inside tunnels or other conveyances where GPS signals are blocked by terrain.

GPS Based

The low price and ubiquity of Global Positioning System or GPS equipment has lent itself to more accurate and reliable telelocation systems. GPS signals are impervious to most electrical noise sources and don't require the user to install an entire system. Only a receiver to collect signals from the satellite segment is installed in each vehicle and a radio to communicate the collected location data with a dispatch point are typically used.

Large private telelocation or automatic vehicle location (AVL) systems send data from GPS receivers in vehicles to a dispatch center over their private, user-owned radio backbone. These systems are used for businesses like parcel delivery and ambulances. Location data is periodically polled from each vehicle in a fleet by a central controller or computer. In the simplest systems, data from the GPS receiver is displayed on a map, allowing people to determine the location of each vehicle. More complex systems feed the data into a computer-assisted dispatch system, which automates the process.

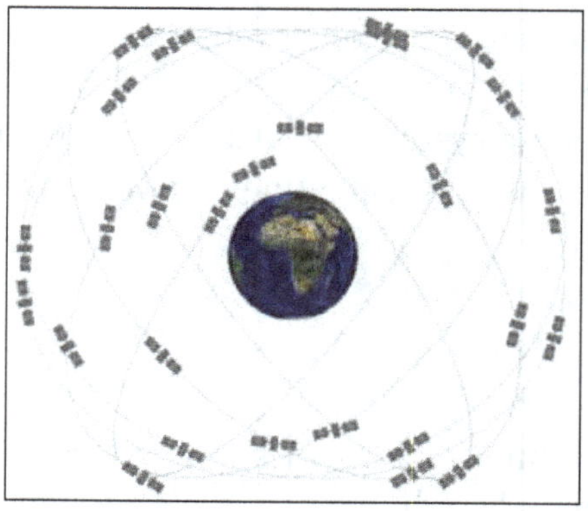

Geographic Coordinate System

A geographic coordinate system is a coordinate systemthat enables every location on Earth to be specified by a set of numbers, letters or symbols. The coordinates are often chosen such that one of the numbers represents a vertical position and two or three of the numbers represent a horizontal position; alternatively, a geographic position may be expressed in a combined three-dimensional Cartesian vector. A common choice of coordinates is latitude, longitude and elevation. To specify a location on a plane requires a map projection.

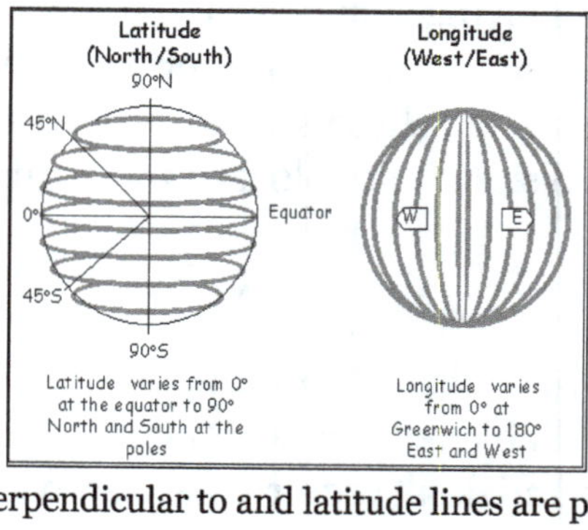

Longitude lines are perpendicular to and latitude lines are parallel to the Equator.

Geodetic Datum

In order to be unambiguous about the direction of "vertical" and the "horizontal" surface above which they are measuring, map-makers choose a reference ellipsoid with a given origin and orientation that best fits their need for the area to be mapped. They then choose the most appropriate mapping of the spherical coordinate system onto that ellipsoid, called a terrestrial reference system or geodetic datum.

Datums may be global, meaning that they represent the whole Earth, or they may be local, meaning that they represent an ellipsoid best-fit to only a portion of the Earth. Points on the Earth's surface move relative to each other due to continental plate motion, subsidence, and diurnal Earth tidal movement caused by the Moon and the Sun. This daily movement can be as much as a meter. Continental movement can be up to 10 cm a year, or 10 m in a century. A weather system high-pressure area can cause a sinking of 5 mm. Scandinavia is rising by 1 cm a year as a result of the melting of the ice sheets of the last ice age, but neighbouring Scotland is rising by only 0.2 cm. These changes are insignificant if a local datum is used, but are statistically significant if a global datum is used.

Examples of global datums include World Geodetic System (WGS 84, also known as EPSG:4326), the default datum used for the Global Positioning System, and the International Terrestrial Reference Frame (ITRF), used for estimating continental drift and crustal deformation. The distance to Earth's center can be used both for very deep positions and for positions in space.

Local datums chosen by a national cartographical organisation include the North American Datum, the European ED50, and the British OSGB36. Given a location, the datum provides the latitude ϕ and longitude λ. In the United Kingdom there are three common latitude, longitude, and height systems in use. WGS 84 differs at Greenwich from the one used on published maps OSGB36 by approximately 112 m. The military system ED50, used by NATO, differs from about 120 m to 180 m.

The latitude and longitude on a map made against a local datum may not be the same as one obtained from a GPS receiver. Converting coordinates from one datum to another requires a datum transformation such as a Helmert transformation, although in certain situations a simple translation may be sufficient.

In popular GIS software, data projected in latitude/longitude is often represented as a *Geographic Coordinate System*. For example, data in latitude/longitude if the datum is the North American Datum of 1983 is denoted by ‹GCS North American 1983'.

Horizontal Coordinates

Latitude and Longitude

The "latitude" (abbreviation: Lat., φ, or phi) of a point on Earth's surface is the angle between the equatorial plane and the straight line that passes through that point and through (or close to) the center of the Earth. Lines joining points of the same latitude trace circles on the surface of Earth called parallels, as they are parallel to the Equator and to each other. The North Poleis 90° N; the South Pole is 90° S. The 0° parallel of latitude is designated the Equator, the fundamental plane of all geographic coordinate systems. The Equator divides the globe into Northern and Southern Hemispheres.

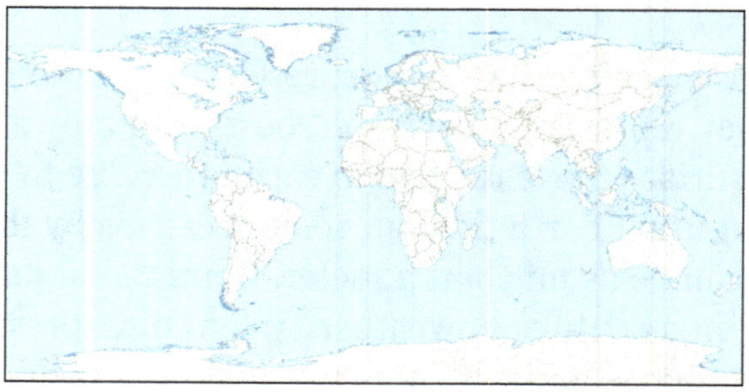

0°

Equator, the 0° parallel of latitude.

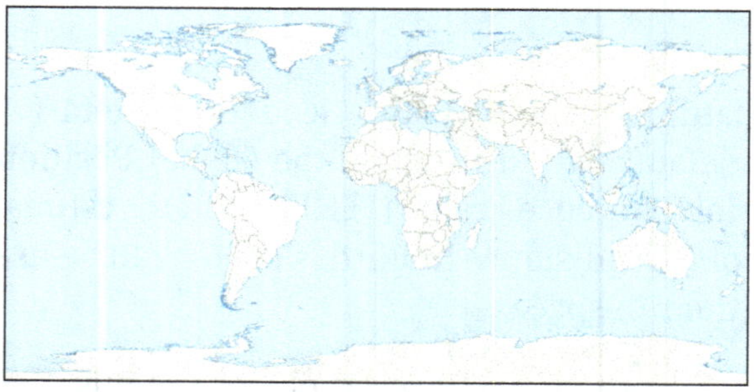

0°

Prime Meridian, the 0° of longitude.

The "longitude" (abbreviation: Long., λ, or lambda) of a point on Earth's surface is the angle east or west of a reference meridian to another meridian that passes through that point. All meridians are halves of great ellipses(often called great circles), which converge at the North and South Poles. The meridian of the British Royal Observatory in Greenwich, in south-east London, England, is the international prime meridian, although some organizations—such as the French Institut Géographique National—continue to use other meridians for internal purposes. The prime meridian determines the proper Eastern and Western Hemispheres, although maps often divide these hemispheres further west in order to keep the Old World on a single side. The antipodal meridian of Greenwich is both 180°W and 180°E. This is not to be conflated with the International Date Line, which diverges from it in several places for political and convenience reasons, including between far eastern Russia and the far western Aleutian Islands.

The combination of these two components specifies the position of any location on the surface of Earth, without consideration of altitude or depth. The grid formed by lines of latitude and longitude is known as a "graticule".The origin/zero point of this system is located in the Gulf of Guinea about 625 km (390 mi) south of Tema, Ghana.

Length of a Degree

On the GRS80 or WGS84 spheroid at sea level at the Equator, one latitudinal second measures 30.715 meters, one latitudinal minute is 1843 meters and one latitudinal degree is 110.6 kilometers. The circles of longitude, meridians, meet at the geographical poles, with the west-east width of a second naturally decreasing as latitude increases. On the Equator at sea level, one longitudinal second measures 30.92 meters, a longitudinal minute is 1855 meters and a longitudinal degree is 111.3 kilometers. At 30° a longitudinal second is 26.76 meters, at Greenwich (51°28′38″N) 19.22 meters, and at 60° it is 15.42 meters.

On the WGS84 spheroid, the length in meters of a degree of latitude at latitude φ (that is, the number of meters you would have to travel along a north–south line to move 1 degree in latitude, when at latitude φ), is about,

$$111132.92 - 559.82\cos 2\varphi + 1.175\cos 4\varphi - 0.0023\cos 6\varphi$$

The returned measure of meters per degree latitude varies continuously with latitude.

Similarly, the length in meters of a degree of longitude can be calculated as,

$$111412.84\cos\varphi - 93.5\cos 3\varphi + 0.118\cos 5\varphi$$

(Those coefficients can be improved, but as they stand the distance they give is correct within a centimeter.)

The formulae both return units of meters per degree.

An alternative method to estimate the length of a longitudinal degree at latitude φ is to assume a spherical Earth (to get the width per minute and second, divide by 60 and 3600, respectively):

$$\frac{\pi}{180} M_r \cos\varphi,$$

where Earth's average meridional radius M_r is 6,367,449 m. Since the Earth is an oblate spheroid, not spherical, that result can be off by several tenths of a percent; a better approximation of a longitudinal degree at latitude φ is:

$$\frac{\pi}{180} a\cos\beta,$$

where, Earth's equatorial radius a equals *6,378,137 m* and $\tan\beta = \frac{b}{a}\tan\varphi$; for the GRS80 and WGS84 spheroids, b/a calculates to be 0.99664719. (β is known as the reduced (or parametric) latitude). Aside from rounding, this is the exact distance along a parallel of latitude; getting the distance along the shortest route will be more work, but those two distances are always within 0.6 meter of each other if the two points are one degree of longitude apart.

Longitudinal length equivalents at selected latitudes					
Latitude	City	Degree	Minute	Second	**±0.0001°**
60°	Saint Petersburg	55.80 km	0.930 km	15.50 m	5.58 m
51° 28′ 38″ N	Greenwich	69.47 km	1.158 km	19.30 m	6.95 m
45°	Bordeaux	78.85 km	1.31 km	21.90 m	7.89 m
30°	New Orleans	96.49 km	1.61 km	26.80 m	9.65 m
0°	Quito	111.3 km	1.855 km	30.92 m	11.13 m

Map Projection

To establish the position of a geographic location on a map, a map projection is used to convert geodetic coordinates to plane coordinates on a map; it projects the datum ellipsoidal coordinates and height onto a flat surface of a map. The datum, along with a map projection applied to a grid of reference locations, establishes a *grid system* for plotting locations. Common map projections in current use include the Universal Transverse Mercator (UTM), the Military Grid Reference System (MGRS), the United States National Grid(USNG), the Global Area Reference System (GARS) and the World Geographic Reference System (GEOREF). Coordinates on a map are usually in terms northing N and easting E offsets relative to a specified origin.

Map projection formulas depend in the geometry of the projection as well as parameters dependent on the particular location at which the map is projected. The set of parameters can vary based on type of project and the conventions chosen for the projection. For the transverse Mercator projectionused in UTM, the parameters associated are the latitude and longitude of the natural origin, the false northing and false easting, and an overall scale factor. Given the parameters associated with particular location or grin, the projection formulas for the transverse Mercator are a complex mix of algebraic and trigonometric functions.

UTM and UPS Systems

The Universal Transverse Mercator (UTM) and Universal Polar Stereographic(UPS) coordinate systems both use a metric-based Cartesian grid laid out on a conformally projected surface to locate positions on the surface of the Earth. The UTM system is not a single map projection but a series of sixty, each covering 6-degree bands of longitude. The UPS system is used for the polar regions, which are not covered by the UTM system.

Stereographic Coordinate System

During medieval times, the stereographic coordinate system was used for navigation purposes. The stereographic coordinate system was superseded by the latitude-longitude system. Although no longer used in navigation, the stereographic coordinate system is still used in modern times to describe crystallographic orientations in the fields of crystallography, mineralogy and materials science.

3D Cartesian Coordinates

Every point that is expressed in ellipsoidal coordinates can be expressed as an rectilinear x y z (Cartesian) coordinate. Cartesian coordinates simplify many mathematical calculations. The Cartesian systems of different datums are not equivalent.

Earth-centered and Earth-fixed

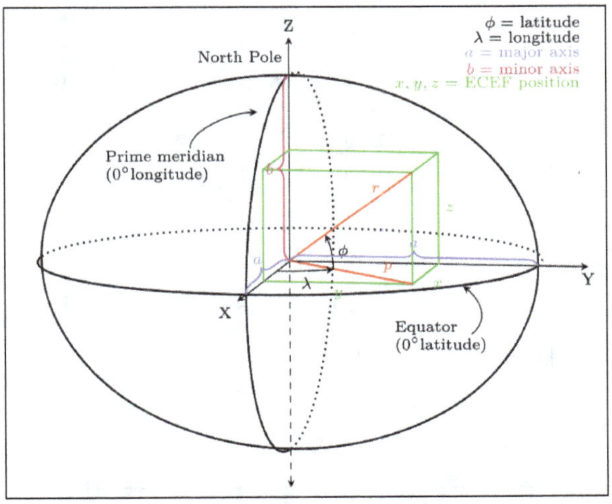

Earth centered, Earth fixed coordinates in relation to latitude and longitude.

The Earth-centered Earth-fixed (also known as the ECEF, ECF, or conventional terrestrial coordinate system) rotates with the Earth and has its origin at the center of the Earth.

The conventional right-handed coordinate system puts:

- The origin at the center of mass of the Earth, a point close to the Earth's center of figure.
- The Z axis on the line between the North and South Poles, with positive values increasing northward (but does not exactly coincide with the Earth's rotational axis).
- The X and Y axes in the plane of the Equator.
- The X axis passing through extending from 180 degrees longitude at the Equator (negative) to 0 degrees longitude (prime meridian) at the Equator (positive).
- The Y axis passing through extending from 90 degrees west longitude at the Equator (negative) to 90 degrees east longitude at the Equator (positive).

An example is the NGS data for a brass disk near Donner Summit, in California. Given the dimensions of the ellipsoid, the conversion from lat/lon/height-above-ellipsoid coordinates to X-Y-Z is straightforward—calculate the X-Y-Z for the given lat-lon on the surface of the ellipsoid and add the X-Y-Z vector that is perpendicular to the ellipsoid there and has length equal to the point's height above the ellipsoid. The reverse conversion is harder: given X-Y-Z we can immediately get longitude, but no closed formula for latitude and height exists. Using Bowring's formula in 1976 *Survey Review* the first iteration gives latitude correct within 10^{-11} degree as long as the point is within 10000 meters above or 5000 meters below the ellipsoid.

Local Tangent Plane

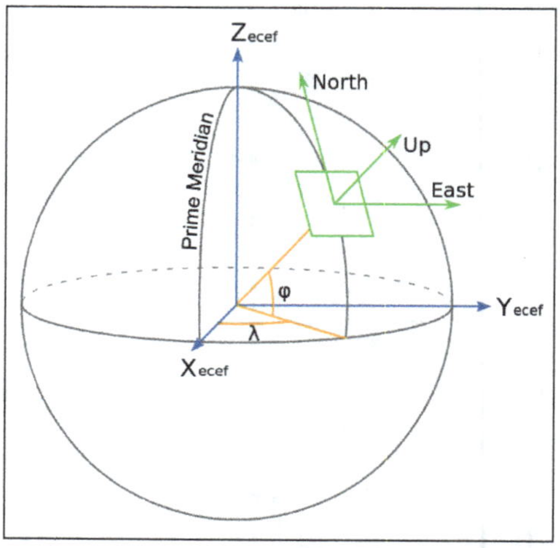

Earth centered Earth fixed and East, North, up coordinates.

A local tangent plane can be defined based on the vertical and horizontaldimensions. The vertical coordinatecan point either up or down. There are two kinds of conventions for the frames:

- East, North, up (ENU), used in geography.
- North, East, down (NED), used specially in aerospace.

In many targeting and tracking applications the local ENU Cartesian coordinate system is far more intuitive and practical than ECEF or geodetic coordinates. The local ENU coordinates are formed from a plane tangent to the Earth's surface fixed to a specific location and hence it is sometimes known as a *local tangent* or *local geodetic* plane. By convention the east axis is labeled x, the north y and the up z.

In an airplane, most objects of interest are below the aircraft, so it is sensible to define down as a positive number. The NED coordinates allow this as an alternative to the ENU. By convention, the north axis is labeled x', the east y' and the down z'. To avoid confusion between x and x', etc. We will restrict the local coordinate frame to ENU.

On other Celestial Bodies

Similar coordinate systems are defined for other celestial bodies such as:

- A similarly well-defined system based on the reference ellipsoid for Mars.
- Selenographic coordinates for the Moon.

Wi-Fi Positioning System

Wi-Fi positioning system (WPS), WiPS or WFPS is a geolocation system that uses the characteristics of nearby Wi-Fi hotspots and other wireless access points to discover where a device is located. It is used where satellite navigation such as GPS is inadequate due to various causes including multipath and signal blockage indoors, or where acquiring a satellite fix would take too long. Such systems include assisted GPS, urban positioning services through hotspot databases, and indoor positioning systems. Wi-Fi positioning takes advantage of the rapid growth in the early 21st century of wireless access points in urban areas.

The most common and widespread localization technique used for positioning with wireless access points is based on measuring the intensity of the received signal (*received signal strength indication* or RSSI) and the method of "fingerprinting". Typical parameters useful to geolocate the wireless access point include its SSID and MAC address. The accuracy depends on the number of nearby access points whose positions have been entered into the database. The Wi-Fi hotspot database gets filled by correlating mobile device GPS location data with Wi-Fi hotspot MAC addresses. The possible signal fluctuations that may occur can increase errors and inaccuracies in the path of the user. To minimize fluctuations in the received signal, there are certain techniques that can be applied to filter the noise.

In the case of low precision, some techniques have been proposed to merge the Wi-Fi traces with other data sources such as geographical information and time constraints (i.e., time geography).

Motivation and Applications

Accurate indoor localization is becoming more important for Wi-Fi based devices due to the increased use of augmented reality, social networking, health care monitoring, personal tracking, inventory control and other indoor location-aware applications.

The popularity and low price of Wi-Fi network interface cards is an attractive incentive to use Wi-Fi as the basis for a localization system and significant research has been done in this area in the past 15 years.

Problem Statement and basic Concepts

The problem of Wi-Fi based indoor localization of a device consists in determining the position of client devices with respect to access points. Many techniques exist to accomplish this, and these may be classified into four main types: *received signal strength indication* (RSSI), fingerprinting, angle of arrival (AoA) and time of flight (ToF) based techniques.

In most cases the first step to determine a device's position is to determine the distance between the target client device and a few access points. With the known distances between the target device and access points, trilaterationalgorithms may be used to determine the relative position of the target device, using the known position of access points as a reference. Alternatively, the angle of arriving signals at a target client device can be employed to determine the device's location based on triangulationalgorithms.

A combination of these techniques may be used to increase the accuracy of the system.

Existing Localization Techniques

RSSI and Lateration Based

RSSI localization techniques are based on measuring signal strength from a client device to several different access points, and then combining this information with a propagation model to determine the distance between the client device and the access points. Trilateration (sometimes called multilateration) techniques can be used to calculate the estimated client device position relative to the known position of access points.

Though one of the cheapest and easiest methods to implement, its disadvantage is that it does not provide very good accuracy (median of 2-4m), because the RSSI measurements tend to fluctuate according to changes in the environment or multipath fading.

Fingerprinting Based

Traditional fingerprinting is also RSSI-based, but it simply relies on the recording of the signal strength from several access points in range and storing this information in a database along with the known coordinates of the client device in an offline phase. This information can be deterministic or probabilistic. During the online tracking phase, the current RSSI vector at an unknown location is compared to those stored in the fingerprint and the closest match is returned as the estimated user location. Such systems may provide a median accuracy of 0.6m and tail accuracy of 1.3m.

Its main disadvantage is that any changes of the environment such as adding or removing furniture or buildings may change the "fingerprint" that corresponds to each location, requiring an update to the fingerprint database. However, the integration with other sensor such as camera can be used in order to deal with changing environment.

Angle of Arrival Based

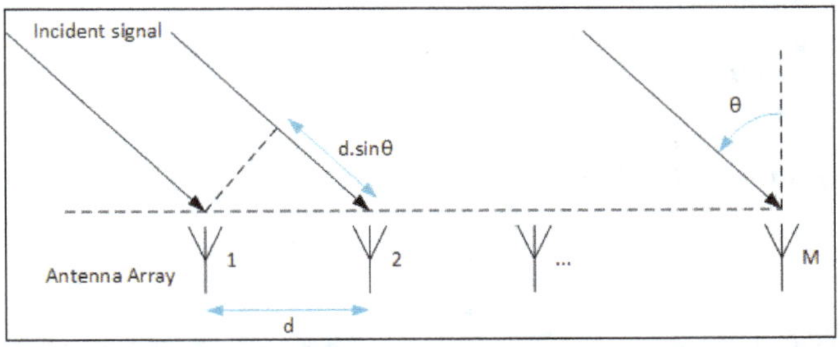

Linear array of antennas receiving a signal. The phase-shift difference of the received signal arriving at antennas equally separated by a "d" distance is used to compute the angle of arrival of the signal.

With the advent of MIMO Wi-Fi interfaces, which use multiple antennas, it is possible to estimate the AoA of the multipath signals received at the antenna arrays in the access points, and apply triangulation to calculate the location of client devices. SpotFi, ArrayTrack and LTEye are proposed solutions which employ this kind of technique.

Typical computation of the AoA is done with the MUSIC algorithm. Assuming an antenna array of M antennas equally spaced by a distance of d and a signal arriving at the antenna array through L propagation paths, an additional distance of $d \cdot sin\theta$ is traveled by the signal to reach the second antenna of the array.

Considering that the kth propagation path arrives with angle θ_k with respect to the normal of the antenna array of the access point, γ_k is the attenuation experienced at any antenna of the array. The attenuation is the same in every antenna, except for a phase shift which changes for every antenna due to the extra distance traveled by the signal. This means that the signal arrives with an additional phase of:

$$-2\pi \cdot d \cdot \sin(\theta) \cdot (f / c) \cdot (2-1),$$

at the second antenna and,

$$-2\pi \cdot d \cdot \sin(\theta) \cdot (f / c) \cdot (m-1),$$

at the mth antenna.

Therefore, the following complex exponential can be used as a simplified representation of the phase shifts experienced by each antenna as a function of the AoA of the propagation path:

$$\phi(\theta_k) = \exp(-j \cdot 2\pi \cdot d \cdot \sin(\theta_k) \cdot f / c)$$

The AoA can then be expressed as the vector $\vec{a}(\theta_k)\gamma_k$ of received signals due to the kth propagation path, where $\vec{a}(\theta_k)$ is the steering vector and given by:

$$\vec{a}(\theta_k) = [1, \phi(\theta_k), ..., \phi(\theta_k)^{M-1}]^T$$

There is one steering vector for each propagation path, and the steering matrix $\mathbf{A}$ (of dimensions $M \bullet L$) is then defined as:

$$\mathbf{A} = [\vec{a}(\theta_1),...,\vec{a}(\theta_L)],$$

and the received signal vector $\vec{x}$ is:

$$\vec{x} = \mathbf{A}\vec{\Gamma},$$

where $\vec{\Gamma} = [\vec{\gamma}_1...\vec{\gamma}_L]$ is the vector complex attenuations along the L paths. OFDM transmits data over multiple different sub carriers, so the measured received signals $\vec{x}$ corresponding to each sub carrier form the matrix $\mathbf{X}$ expressed as:

$$\mathbf{X} = [\vec{x}_1...\vec{x}_L] = \mathbf{A}[\vec{\Gamma}_1...\vec{\Gamma}_L] = \mathbf{AF}$$

The matrix $\mathbf{X}$ is given by the channel state information (CSI) matrix which can be extracted from modern wireless cards with special tools such as the Linux 802.11n CSI Tool.

This is where the MUSIC algorithm is applied in, first by computing the eigenvectors of $\mathbf{XX}^H$ (where $\mathbf{X}^H$ is the conjugate transpose of $\mathbf{X}$) and using the vectors corresponding to eigenvalue zero to calculate the steering vectors and the matrix $\mathbf{A}$. The AoAs can then be deduced from this matrix and used to estimate the position of the client device through triangulation.

Though this technique is usually more accurate than others, it may require special hardware in order to be deployed, such as an array of six to eight antennas or rotating antennas. SpotFi proposes the use of a superresolution algorithm which takes advantage of the number of measurements taken by each of the antennas of the Wi-Fi cards with only three antennas, and also incorporates ToF based localization to improve its accuracy.

Time of Flight Based

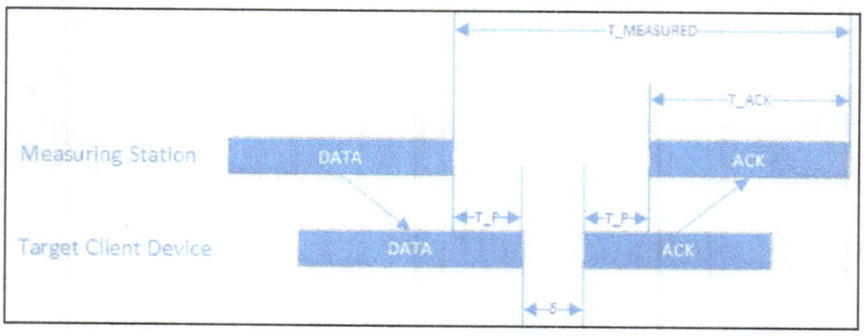

Figure showing a measuring station sending a DATA frame to a client station and waiting until receiving the ACK. δ is the scheduling delay (offset) originated at the target client device, and it depends on how much time it takes for the ACK to be scheduled.

T_P is the signal propagation time between transmitter and receiver, and is usually assumed to be the same on the way to the target and back. T_ACK is the time needed to transmit the ACK frame. The time of flight corresponds to the T_MEASURED. Picture reproduced from

This localization approach takes timestamps provided by the wireless interfaces to calculate the ToF of signals and then use this information to estimate the distance and relative position of one client device with respect to access points. The granularity of such time measurements is in the order of nanoseconds and systems which use this technique have reported localization errors in the order of 2m. Typical applications for this technology are tagging and locating assets in buildings, for which room-level accuracy (~3m) is usually enough.

The time measurements taken at the wireless interfaces are based on the fact that RF waves travel close to the speed of light, which remains nearly constant in most propagation media in indoor environments. Therefore, the signal propagation speed (and consequently the ToF) is not affected so much by the environment as the RSSI measurements are.

Unlike traditional ToF-based echo techniques, such as those used in RADARsystems, Wi-Fi echo techniques use regular data and acknowledgement communication frames to measure the ToF.

As in the RSSI approach, the ToF is used only to estimate the distance between the client device and access points. Then a trilateration technique can be used to calculate the estimated position of the device relative to the access points. The greatest challenges in the ToF approach consist in dealing with clock synchronization issues, noise, sampling artifacts and multipath channel effects. Some techniques uses mathematical approaches to remove the need for clock synchronization.

Ubiquitous Indoor Localization on a World-wide Scale

Recently, researchers have envisioned an Indoor Positioning System (IPS) that can perform ubiquitous indoor localization on a worldwide scale, similar to the GPS outdoors. The idea is to leverage the sensors embedded in today's smartphones in a crowd-sourcing manner to automatically construct the indoor floorplans as well as automatically construct the Wi-Fi fingerprint. This not only automatically builds the fingerprints on a worldwide scale, but also keeps the constructed fingerprints up-to-date.

Privacy Concerns

Citing the specific privacy concerns arising out of WPS, Google suggested a unified approach for opting-out a particular access point from taking part in determining location using WPS. Appending "_nomap" to a wireless access point's SSID excludes it from Google's WPS database. Google hopes that other WPS providers and data collectors, like Apple and Microsoft, follow that recommendation so that it becomes an accepted standard. Mozillahonors _nomap as a method of opting-out of its location service.

Geolocation Software

In computing, geolocation software is software that is capable of deducing the geolocation of a device connected to the Internet. The identification of a device's IP address can be used to determine the country, city, or post/ZIP code, determining an object's geographical location. Other methods include examination of a MAC address, image metadata, or credit card information.

Data Sources

An IP address is assigned to each device (e.g., computer, printer) participating in a computer network that uses the Internet Protocol for communication. The protocol specifies that each IP packet must have a header which contains, among other things, the IP address of the sender of the packet.

There are a number of free and paid subscription geolocation databases, ranging from country level to state or city—including ZIP/post code level—each with varying claims of accuracy (generally higher at the country level). These databases typically contain IP address data which may be used in firewalls, ad servers, routing, mail systems, web sites, and other automated systems where geolocation may be useful. An alternative to hosting and querying a database is to obtain the country code for a given IP address through a DNSBL-style lookup from a remote server.

Some commercial databases have augmented geolocation software with demographic data to enable demographic-type targeting using IP address data.

The primary source for IP address data is the regional Internet registrieswhich allocate and distribute IP addresses amongst organizations located in their respective service regions:

- African Network Information Centre (AfriNIC).
- American Registry for Internet Numbers (ARIN).
- Asia-Pacific Network Information Centre (APNIC).
- Latin American and Caribbean Internet Address Registry (LACNIC).
- RIPE Network Coordination Centre (RIPE NCC).

Secondary sources include:

- Data mining or user-submitted geographic location data, e.g. a weather website asking visitors for a city name to find their local forecast or pairing a user's IP address with the address information in his/her account profile.

- Data contributed by Internet service providers.
- Merging databases from different suppliers.
- Guesstimates from adjacent Class C range and/or gleaned from network hops.
- Network routing information collected to the end point of IP address.

Accuracy is improved by:

- Data scrubbing to filter out or identify anomalies.
- Statistical analysis of user submitted data.
- Utilizing third-party tests conducted by reputable organizations.

Errors

If geolocation software maps IP addresses associated with an entire county or territory to a particular location (such as the geographic center of the territory), this can cause considerable problems for the people who happen to live there, as law enforcement authorities and others may mistakenly assume any crimes or other misconduct associated with the IP address to originate from that particular location.

For example, a farmstead northeast of Potwin, Kansas became the default site of 600 million IP addresses (due to their lack of fine granularity) when the Massachusetts-based digital mapping company MaxMind changed the putative geographic center of the contiguous United States from 39.8333333,-98.585522 to 38.0000,-97.0000. Similarly, a family in Pretoria, the capital of South Africa, was since 2012 regularly visited by police or angry private citizens who believed their stolen phones were to be found in the family's backyard. This was also the result of geolocation by MaxMind. The company used the National Geospatial-Intelligence Agency's coordinates for Pretoria, which pointed to the family's house, to represent IP addresses associated with Pretoria.

Privacy

A distinction can be made between *co-operative* and *oppositional* geolocation. In some cases, it is in the interest of users to be accurately located, for example, so that they can be offered information relevant to their location. In other cases, users prefer to not disclose their location for privacy or other reasons.

Technical measures for ensuring anonymity, such as proxy servers, can be used to circumvent restrictions imposed by geolocation software. Some sites detect the use of proxies and anonymizers, and may either block service or provide non-localized content in response.

Applications

Geolocation technology has been under development only since 1999, and the first patents were granted in 2004. The technology is already widely used in multiple industries, including e-retail, banking, media, telecommunications, education, travel, hospitality, entertainment, health care, online gaming and law enforcement, for preventing online fraud, complying with regulations, managing digital rights and serving targeted marketing content and pricing. Additionally, the U.S. Federal Communications Commission (FCC) has proposed that geolocation software might be leveraged to support 9-1-1 location determination.

An IP address or the related unique URL may also investigated with basic functions, typing from the keyboard two instructions: ping and traceroute.In Unix-like systems, they are available as a command line tool. In the same way, Microsoft Windows has the prompt of DOS working with those instructions.

Criminal Investigations

Banks, software vendors and other online enterprises in the USA and elsewhere became subject to strict "know your customer" laws imposed by the USA PATRIOT Act, the Bank Secrecy Act, the US Treasury Department's Office of Foreign Assets Control and other regulatory entities in the US and Europe from the early twenty-first century. These laws are intended to prevent money laundering, trafficking with terrorist organizations, and trading with banned nations. When it is possible to identify the true location of online visitors, geolocation can protect banks from participating in the transfer of funds for illicit purposes. More and more prosecuting bodies are bringing cases involving cyber-crimes such as cyber-stalking and identity theft. Prosecutors often have the capability of determining the IP address data necessary to link a computer to a crime.

Fraud Detection

Online retailers and payment processors use geolocation to detect possible credit card fraud by comparing the user's location to the billing address on the account or the shipping address provided. A mismatch an order placed from the US on an account number from Tokyo, for example is a strong indicator of potential fraud. IP address geolocation can be also used in fraud detectionto match billing address postal code or area code. Banks can prevent "phishing" attacks, money laundering and other security breaches by determining the user's location as part of the authentication process. Whoisdatabases can also help verify IP addresses and registrants.

Government, law enforcement and corporate security teams use geolocation as an investigatory tool, tracking the Internet routes of online attackers to find the perpetrators and prevent future attacks from the same location.

Geo Marketing

Since geolocation software can get the information of user location, companies using geomarketing may provide web content or products that are famous or useful in that specific location. Advertisements and content on a website that uses geolocation software in the form of an API (also referred to as "IP API" or "IP address geolocation API") may be tailored to provide the information that a certain user wants.

Regional Licensing

Internet movie vendors, online broadcasters who serve live streaming video of sporting events, or certain TV and music video sites that are licensed to broadcast their videos of episodes/music videos are permitted to serve viewers only in their licensed territories. By geolocating viewers, they can be certain of obeying licensing regulations. Online gambling websites must also know where their customers violate gambling laws, or risk doing so.

Jim Ramo, chief executive of movie distributor Movielink, said studios were aware of the shortcomings going in and have grown more confident now that the system has been shown to work.

Gaming

A location-based game is a type of pervasive game for smartphones or other mobile devices in which the gameplay evolves and progresses via a player's real-world location which is typically reported via GPS.

Mobile Local Search

Mobile local search is a technology that lets people search for local things using mobile equipment such as mobile phones, PDAs, and other mobile devices. Mobile local search satisfies the need to offer a mobile subscriber spontaneous access to near-position services and information such as businesses, products, events, restaurant, movie theatre or other local information. Mobile local search is the search and discovery of persons, places, and things within an identifiable space defined by distinct parameters. These parameters are evolving. Today they include social networks, individuals, cities, neighbourhoods, landmarks, and actions that are relevant to the searcher's past, current, and future location. These parameters provide structure to vertically deep and horizontally broad data categories that can stand-alone or are combined to comprise searchable directories.

Mobile local search is usually based on organized directories accessed through specialized search tools, rather than the web, although mobile local search often provides links to mobile (WAP) web sites. It is also an application of a location-based service.

Technology

Mobile local search involves in some cases GPS tracking, which, without prior consent, may cause privacy concerns. Furthermore, recent surveys expect merely 10% of mobile devices worldwide to be equipped with GPS chips in 2011.

Alternatively, mobile local search can also be done without mobile operators and GPS systems involved. This is possible through access to the SS7protocol, which mobile operators have. However, having access via the mobile operator is very hard since the mobile operators sell it at a premium rate, if at all.

Although GPS-enabled devices are not prevalent yet, industry experts show that consumers are demanding mobile local search now, with an expected growth of 91% from 2007 to 2011. As a consequence, there is a gap in this opportunity that can be cleverly addressed by mobile operators and service providers with access to the SS7 layer.

Access Methods

Mobile local search may be screen-based using the keypad and display on the mobile device, or voice-based using spoken commands that are interpreted by a speech recognition application.

Screen-based search using the keypad to enter search terms into a form on the mobile screen is the most common access method today. It has the advantages of working in a noisy environment and avoiding speech interpretation problems. But the user has to enter search terms using a small keypad typical of most mobile devices, with results displayed on a small screen. Screen-based search may be awkward for people with coordination or vision handicaps.

Screen-based search may be supported by a specialized application such as where.com running on the handset, by an SMS short code, or by a WAP or XHTML-MP website page on a remote server displayed by the handset browser. Although most search tools today are based on handset software, the trend seems to be towards web-based services.

Also, when the consumer receives a reply to his search request, the most common way of delivery is via SMS.

Voice-based search is becoming viable due to advances in speech recognition technology. It is useful for people who have difficulty using the small keypad, or who need to get information while driving. Spoken prompts guide the users through the local directory to find an entry. The service may include fallback to a live operator if the speech technology has difficulty understanding what they are looking for, or a 'silent agent' may assist the speech recognition software.

Results may be spoken and/or delivered to the handset in SMS or MMSformat, where they may be stored for future reference.

Most often, voice-based search uses the voice channel to deliver the caller's utterance to a server-based search application. In some cases, a small application on the mobile device undertakes the first part of speech recognition, breaking down the utterance into a series of elementary components or 'phones', and then sending the phones over the data channel to a server application.

Multimedia-based search will be prevalent as a hybrid form of both "voice-based" and "screen-based" search that will establish a time-memory tradeoff in the implementation process of fast-evolving mobile equipment.

Types of Information

Mobile search content may include more location-based and more street-smart information on businesses, products, services, events, human relations and other local specifics,which require immediacy and spontaneity. Besides, the location-based content may be formulated by accumulative data mining on consumer behaviors and whereabouts, which may trigger privacy concerns,commercial come-ons and, if necessary, some legislative justifications.

A search for businesses is the mobile equivalent of the printed Yellow Pages, but the search may focus on a small geographical area. For example, the caller may be looking for a nearby pharmacy or restaurant, or may want a local pizza delivery. The more advanced local search tools will vary the search radius depending on the type of business and the density of this type of business in the local area. For example, a caller looking for a coffee shop will probably want one within walking distance. If they are looking for a home furniture store they will expect to travel further. If they want a taxi, distance is not important: the taxi will come to them.

A search for products is more specific than a search for businesses. The user is looking for a particular product and wants to find a local business that stocks this product.

A full-function mobile local search service will go much further, particularly as smartphone penetration increases. For example, mobile users may want to find where a particular movie is playing, what restaurants are near the cinema, read reviews about those restaurants, and book reservations or buy movie tickets. They may want to hear about traffic conditions, the score from the local football game, the winning lottery number or where to go to vote.

Differences from Web Search

As opposed to Web search, mobile local search requires more immediacy and street-smart knowledge and information, local events and driving directions. The technical difference from web-based search is that, whether using the keypad or voice commands, the caller has limited 'bandwidth'. They do not want to hear or scroll through long lists of results. They want to "quickly and unobtrusively" zero in on the type of local information they are looking for, due to privacy, safety and time pressure.

Some of the advancements by the major portals in Internet search, such as Google's famous page-ranking scheme, do not apply in the wireless world since people are not searching for Web sites as much as answers to specific questions. 2006 showed a significant movement to the question-answer model. In this model answers are sent in reply to directory service queries similar to the nature of conventional 411 operators. AskMeNow, Ask.com and Bing.com all made efforts in this regard.

Business Models

Mobile local search services are usually provided today by mobile carriers, directory enquiry providers, mobile messaging operators or Yellow Pages publishers. Behind these outlets, there is a growing network of agencies, distributors and software providers, some of whom are now starting to offer service direct to the public. Search giants, like Yahoo and Google, and big web publishers, such as Amazon.con and eBay have already embarked on this potentially lucrative market segment.

The services are provided through a caller-pays, advertiser-pays or hybrid business model.

In a caller-pays model, as with voice-based directory enquiry services, the caller pays each time they access the service. Caller-pays voice-based local mobile search services are common outside North America, but have been slow to emerge in the USA due to regulatory restrictions.

In an advertiser-pays model, standard for screen-based search and starting to emerge with voice-based ‹Free DA' services in North America, businesses pay to be placed early in the list of results, or pay each time a caller chooses to connect through.

In hybrid models, advertising revenue lets the service provider offer reduced rates.

As with TV, magazines and other media, mobile local search services will most likely continue to be offered under all three models, but the hybrid model may predominate. That is, callers will probably accept a limited amount of advertising (which they may actually find useful) in exchange for reduced prices, but may be less willing to use a free service where they are bombarded with advertising.

The value to an advertiser will vary depending on their type of business, and may range from a dollar or less for a taxi or sandwich shop to well over ten dollars for a real-estate broker, attorney or debt consolidation service. Consultants forecast that the global industry will grow to over US$1 billion by 2010.

Service Providers

The most direct way to find more about what is happening in the MoLo space is to check out the websites of the MoLo service providers and see what they are offering. There are many of them. Almost all mobile carriers have mobile local search offerings, including Verizon Wireless, AT&T, Vodafone, T-Mobile, Sprint Nextel, Orange SA and so on.

Many Directory Assistance providers also offer mobile local search, including free DA providers Jingle Networks and 1-800-San-Diego in the US, 118 118, Yell Group, Pages Jaunes, Seat Pagine Gialle, Eniro etc. in Europe, and many others. One of the more fully functional services is provided by Contact Center Americas in Colombia.

The Yellow Pages Group in Canada provides one of the first voice-enabled local search services.

Mobile Phone Tracking

Mobile phone tracking is a process for identifying the location of a mobile phone, whether stationary or moving. Localization may be effected by a number of technologies, such as using multilateration of radio signals between (several) cell towers of the network and the phone, or simply using GPS. To locate a mobile phone using multilateration of radio signals, it must emit at least the roamingsignal to contact the next nearby antenna tower, but the process does not require an active call. The Global System for Mobile Communications (GSM) is based on the phone's signal strength to nearby antenna masts.

Mobile positioning may include location-based services that disclose the actual coordinates of a mobile phone, which is a technology used by telecommunication companies to approximate the location of a mobile phone, and thereby also its user.

Technology

The location of a mobile phone can be determined in a number of ways.

Network-based

The location of a mobile phone can be determined using the service provider's network infrastructure. The advantage of network-based techniques, from a service provider's point of view, is that they can be implemented non-intrusively without affecting handsets. Network-based techniques were developed many years prior to the widespread availability of GPS on handsets.

The technology of locating is based on measuring power levels and antenna patterns and uses the concept that a powered mobile phone always communicates wirelessly with one of the closest base stations, so knowledge of the location of the base station implies the cell phone is nearby.

Advanced systems determine the sector in which the mobile phone is located and roughly estimate also the distance to the base station. Further approximation can be done by interpolating signals between adjacent antenna towers. Qualified services may achieve a precision of down to 50 meters in urban areas where mobile traffic and density of antenna towers (base stations) is sufficiently high. Rural and desolate areas may see miles between base stations and therefore determine locations less precisely.

GSM localization uses multilateration to determine the location of GSM mobile phones, or dedicated trackers, usually with the intent to locate the user.

The accuracy of network-based techniques varies, with cell identification as the least accurate (due to differential signals transposing between towers, otherwise known as "bouncing signals") and triangulation as moderately accurate, and newer "advanced forward link trilateration" timing methods as the most accurate. The accuracy of network-based techniques is both dependent on the concentration of cell base stations, with urban environments achieving the highest possible accuracy because of the higher number of cell towers, and the implementation of the most current timing methods.

One of the key challenges of network-based techniques is the requirement to work closely with the service provider, as it entails the installation of hardware and software within the operator's infrastructure. Frequently the compulsion associated with a legislative framework, such as Enhanced 9-1-1, is required before a service provider will deploy a solution.

Handset-based

The location of a mobile phone can be determined using client softwareinstalled on the handset. This technique determines the location of the handset by putting its location by cell identification, signal strengths of the home and neighboring cells, which is continuously sent to the carrier. In addition, if the handset is also equipped with GPS then significantly more precise location information can be then sent from the handset to the carrier.

Another approach is to use a fingerprinting-based technique, where the "signature" of the home and neighboring cells signal strengths at different points in the area of interest is recorded by war-driving and matched in real-time to determine the handset location. This is usually performed independent from the carrier.

The key disadvantage of handset-based techniques, from service provider's point of view, is the necessity of installing software on the handset. It requires the active cooperation of the mobile subscriber as well as software that must be able to handle the different operating systems of the handsets. Typically, smartphones, such as one based on Symbian, Windows Mobile, Windows Phone, BlackBerry OS, iOS, or Android, would be able to run such software, e.g. Google Maps.

One proposed work-around is the installation of embedded hardware or software on the handset by the manufacturers, e.g., Enhanced Observed Time Difference (E-OTD).

This avenue has not made significant headway, due to the difficulty of convincing different manufacturers to cooperate on a common mechanism and to address the cost issue. Another difficulty would be to address the issue of foreign handsets that are roaming in the network.

SIM-based

Using the subscriber identity module (SIM) in GSM and Universal Mobile Telecommunications System (UMTS) handsets, it is possible to obtain raw radio measurements from the handset. Available measurements include the serving Cell ID, round-trip time, and signal strength. The type of information obtained via the SIM can differ from that which is available from the handset. For example, it may not be possible to obtain any raw measurements from the handset directly, yet still obtain measurements via the SIM.

Wi-Fi

Crowdsourced Wi-Fi data can also be used to identify a handset's location.The poor performance of the GPS-based methods in indoor environment and the increasing popularity of Wi-Fi have encouraged companies to design new and feasible methods to carry out Wi-Fi-based indoor positioning. Most smartphones combine Global Navigation Satellite Systems (GNSS), such as GPS and GLONASS, with Wi-Fi positioning systems.

Hybrid Positioning System

Hybrid positioning systems use a combination of network-based and handset-based technologies for location determination. One example would be some modes of Assisted GPS, which can both use GPS and network information to compute the location. Both types of data are thus used by the telephone to make the location more accurate (i.e., A-GPS). Alternatively tracking with both systems can also occur by having the phone attain its GPS-location directly from the satellites, and then having the information sent via the network to the person that is trying to locate the telephone. Such systems include Google Maps, as well as, LTE's OTDOA and E-CellID.

There are also hybrid positioning systems which combine several different location approaches to position mobile devices by Wi-Fi, WiMAX, GSM, LTE, IP addresses, and network environment data.

Operational Purpose

In order to route calls to a phone, the cell towers listen for a signal sent from the phone and negotiate which tower is best able to communicate with the phone. As the phone changes location, the antenna towers monitor the signal, and the phone is "roamed" to an adjacent tower as appropriate. By comparing the relative signal strength from multiple antenna towers, a general location of a phone can be roughly determined.

Other means make use of the antenna pattern, which supports angular determination and phase discrimination.

Newer phones may also allow the tracking of the phone even when turned on and not active in a telephone call. This results from the roaming procedures that perform handover of the phone from one base station to another.

Consumer Applications

A phone's location can be shared with friends and family, posted to a public web site, recorded locally, or shared with other users of a smartphone app. The inclusion of GPS receivers on smartphones has made geographical apps nearly ubiquitous on these devices. Specific applications include:

- GPS navigation and maps,
- Locator apps like Find My Friends,
- Dating apps like Grindr,
- Recording a journey, for example to show a hiking accomplishment,
- For quantified self purposes such as fitness tracking,
- GPS drawing.

In January 2019, the location of her iPhone as determined by her sister helped Boston police find kidnapping victim Olivia Ambrose.

Indoor Positioning System

Indoor positioning systems (IPS) enable locating the position of objects or people within buildings. Since GPS is unreliable in interior spaces because there is no visual contact with the GPS satellites, an IPS (indoor positioning system) must use other positioning methods. These include, for example, the common consumer standards Wi-Fi or Bluetooth Low Energy (BLE), but also solutions based on Ultra-wideband (UWB) or passive RFID. infsoft's systems are suitable for both indoor and outdoor use. The first and most important step in the implementation of locating systems is the selection of the positioning method and technology.

Working of Indoor Positioning System

When it comes to localization within buildings, a distinction can be made between client-based and server-based positioning. Client-based localization enables determining the position directly on the end user's device (e. g. smartphone). In the case of server-based localization, positioning takes place on a server.

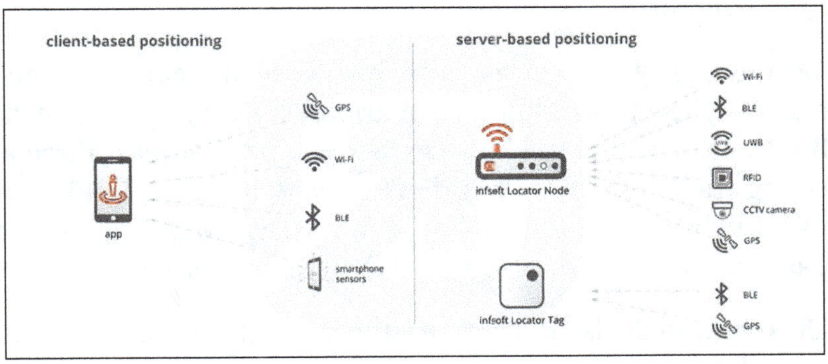

5 Types of Indoor Tracking Technology

Proximity-based Systems

Proximity-based systems can detect the general location of a person or object at room level within a facility. (That's in contrast to a precision system like ultra wide-band, named below, which pinpoints the exact, precise location of something down to a "dot on the map".) Proximity-based systems use tags and beacons for indoor positioning, and they are either reader-based or reference point-based.

In a reader-based system, simple, inexpensive tags ("dumb" tags) transmit their identification continually to a number of reader devices. Those reader devices then pass the tags' identifications and signal strength to a backend system, which then calculates the positions of each tag.

A reference point-based system—of which AirFinder is a good example—uses standard Bluetooth Low Energy (BLE) beacons as location reference points together with "smart" location-aware tags. The tags calculate their own location based on the location of the reference points, and then connect to a central access point to relay this information. The BLE access points, spaced about every 100 feet in a facility, receive the encrypted location data from the tag and send it to the server.

Both types of proximity-based systems can accomplish simple indoor localization at the lowest cost. They are ubiquitous in healthcare and manufacturing, as well as some other industries. Reference point-based systems, however, are the most inexpensive of the proximity solutions because of their architecture—they don't need as many connected readers because of their low-cost reference points. They also enable prolonged battery life and deliver more accurate location information than reader-based systems.

WiFi-based Systems

In a WiFi-based system, tags are WiFi transmitters that send simple packets to a number of WiFi access points in a facility. These access points report the time and strength of that reading to a backend, which uses algorithms to compute position. The location information is then sent to the cloud.

WiFi indoor positioning systems give a fairly high level of accuracy—from three to five meters—because they use time difference of arrival (TDOA) measurements with wide bandwidth. But to achieve this level of accuracy you need at least three access points to "hear" each tag transmission. If you don't already have the WiFi access points to support it, this could be an expensive solution. WiFi tags are also fairly expensive ($40-$60 per tag), and usually less power-efficient than their alternatives.

That said, Wifi is commonly used in both healthcare and manufacturing settings.

Ultra Wide-band Systems

UWB is a cool technology. Three or more ultra wide-band readers transmit a very wide pulse over a GHz of spectrum. The readers then listen for chirps from ultra wide-band tags. These tags have a spark-gap-style exciter that generates a little pulse within them, which creates a short, coded, very wide, nearly instantaneous burst. The readers then report very accurate time measurements from the tags back to a central server.

Because the UWB signal is extremely wide, the accuracy of the location information is very good—probably the most accurate of any system available. A drawback, however, is that UWB is the most expensive system to install. Even though UWB tags are inexpensive, every location has to have at least three (expensive) readers in it because of the limited range of the tags.

Acoustic Systems

A number of new indoor positioning systems have come onto the market that uses ultrasonic pulses from tags to locate them within an indoor environment. An acoustic system works almost exactly like UWB except it uses sound instead of radio. The tags emit a sound in the ultrasonic range (so you can't hear it). Receivers in the room (sometimes multiple, and sometimes a single "smart" one) pick up those sounds and locate the tags that way.

One benefit of using sound has to do with resolving multipath. If you're sending a transmission and taking a time measurement, you can guess the location based on the speed. If that signal bounces off the wall on the way there, you now have a "multipath"—or maybe dozens of them. The ability to mathematically differentiate between a direct path and a multipath is purely a function of the speed of the medium divided by the bandwidth. So your multipath resolution ability is the speed of sound divided by the bandwidth. Acoustic systems require less signal bandwidth to resolve multipath because the speed of sound is so much less than the speed of light.

Sonar-based systems can be very accurate as well—even as accurate as UWB. The cost depends on your situation: If you're installing it in new construction, it won't cost much to wire a sensor into every room. If you're trying to retrofit an existing building with sensors, the cost would be high. Tags, however, are inexpensive.

For now, acoustic systems are an uncommonly-used, niche technology, though health-care offers the most potential for use cases in the near future.

Infrared Systems

Infrared-based indoor localization systems use infrared light pulses (like a TV remote) to locate signals inside a building. IR receivers are installed in every room, and when the IR tag pulses, it is read by the IR receiver device.

Infrared is a near foolproof way to guarantee room-level accuracy. It uses light instead of radio waves, which can't go through walls—if the system says an asset is in room 4B, it is in room 4B without a doubt. Radio-based systems have more trouble with false positives, as the radio waves can sometimes be picked up by other readers through walls.

While the tags are low-cost and long-lasting, a drawback of infrared is that every room needs a wired IR reader to be installed in the ceiling. That's fine if you're installing it in new construction, but, just as with acoustic, retrofitting will be expensive. That's why infrared systems are commonly used in new hospital construction, where rooms are definitively segmented. In an open-space warehouse, infrared would be a challenge—if three receivers read a light pulse, there's no way to know which receiver the tag is closest to because it's difficult to measure the relative signal strength of infrared. Generally speaking, radio technologies work better in open spaces characteristic of warehouses and manufacturing facilities.

Satellite Geolocation

Satellite geolocation is the process of locating the origin of a signal appearing on a satellite communication channel. Typically, this process is used to mitigate interference on communication satellites. Usually, these interference signals are caused by human error or equipment failure, but can also be caused by deliberate jamming. Identifying the geographical location of an interfering signal informs the mitigation activity.

Working of Satellite Geolocation System

Many communication satellites share a given frequency band. As a signal is transmitted to a particular satellite there is some amount of side lobe or spillover energy that is transmitted to adjacent satellites. At a receive station that has two antennas, one pointed at the primary satellite (the satellite the signal is intended for) and a secondary satellite (a satellite that is receiving side lobe energy), both paths of the signal are received and measured. From a comparison of those paths, two measurements can be made: Differential Time Offset (DTO) and Differential Frequency Offset (DFO). These measurements are often implemented through correlation processing. DTO represents the difference in time that it takes the signal to travel through the two satellites, while DFO represents the difference in frequency the received signals present through the two satellites. The frequency differences observed are due to different Doppler shift resulting from relative satellite motion and differences in the translation frequencies of the two satellite channels. Channel translation frequencies and downlink Doppler shift and delay can be calibrated out of the measurements by observing transmitters of known location simultaneously on the channels. This leaves the uplink DTO and DFO as the observables.

Lines of Position

Once a DTO calculated, it can be combined with the known position of the satellites and the receiving station. This combination provides a locus of positions on the Earth's surface for the source of the signal; from this result a line of position (LOP) can be derived. A similar line can be derived for the frequency differences. Where the two LOPs intersect is the signal transmission location. In addition to geolocation with a time LOP and a frequency LOP, a location can also be determined by finding the crossing point of two time LOPs. The second time LOP is an identical measurement using a different secondary satellite, or using the same secondary satellite, but later in time. Similarly, two frequency LOPs can be used to determine a location. It can be shown that, in general, it is expected that the two LOPs intersect in two places. In many circumstances it is possible to discount one of the intersections e.g. due to it not being in the coverage area of one or both satellites. In some circumstances, it is not possible to distinguish intersections from a pair of LOPs, in which case, additional LOPs need to be determined.

Reference Signals

While measuring the DTO and DFO will give you an idea of the location of the signal source, the location will be inaccurate. There are many biases within the measurement system that, if not accounted for properly, will manifest themselves as time delays or frequency offsets. For example, while a satellite translation frequency is known to within a few kHz, accurate geolocation requires frequency measurement accuracies of single mHz.

In order to determine the position of signal source, a second set of measurements is required. Typically, this is done by making DTO and DFO measurements for a reference signal simultaneous with the target signal measurement. The measurement of the reference signal is purely passive and simply serves to remove the biases in the system. The same measurements that are made for the target signal, DTO and DFO, are made for the reference signal. The key to a reference signal is that the transmit location of that signal is known. By comparing the DTO of the reference signal and the DTO of the target signal a result known as Time Difference of Arrival (TDOA) can be calculated. Likewise, from the DFO of the target and the DFO of the reference signal, a Frequency Difference of Arrival (FDOA) can be determined. The TDOA and FDOA results provide a finite number of locations on the Earth's surface, and therefore, lines of position (LOPs) are determined from the TDOA and FDOA results.

A limitation as to how accurately a location can be obtained is knowledge of the satellites' positions and velocities generated by the satellite ephemerides (orbit descriptors). A single reference geographically close to the target will give a high degree of cancellation of the location effects of ephemeris error. Measurements on signals from multiple reference sites can be used to improve the accuracy of the satellite ephemerides thereby provided improved geolocation accuracy generally.

Various Geolocation Methods

TDOA and FDOA results can be gathered and combined in various methods to produce geolocation results. Each method has its advantages and disadvantages in different measurement scenarios.

TDOA-TDOA Geolocation

TDOA-TDOA geolocation is performed, generally, by measuring DTO values using two secondary satellites, or three total satellites. By doing this, two TDOA lines are generated, hopefully, with a crossing point. TDOA – TDOA geolocation is ideal for moving targets, since the movement of the target will introduce varying and random frequency changes, causing an FDOA result to be useless, unless obtained from a highly inclined satellite. TDOA-TDOAgeolocation will not work for unmodulated signals. Due to the repetitive nature of the signal, no unique TDOA solution will exist. One problem with using only TDOA lines of position is that they tend to be north-south orientated and close to parallel, so that the "crossing point" of a TDOA-TDOA measurement can be error prone and uncertain, as it is "hidden" in a long intersection of the lines. Care is also necessary in interpreting the results from moving targets if the two TDOA observations are not obtained simultaneously since the target will have moved between observations.

FDOA-FDOA Geolocation

FDOA-FDOA geolocation is accomplished by using three satellites, or by using time separated measurements on two satellites. The time separation can be as little as 5 minutes or as much as an hour or more. Again, the two FDOA lines are used to find a crossing point, or target location. FDOA-FDOAgeolocation is necessary for CW signals. Geolocation on highly inclined satellites, either one of both being used in the measurement, will result in more accurate results by performing FDOA – FDOA geolocation. This is due to a large difference in relative motion, leading to a large difference in relative frequency between the two satellites. A related point is the error to FDOA-FDOA calculation contributed by ephemeris uncertainty is relatively small. Moving targets are not likely to be located using FDOA methods, unless using a highly inclined satellite. FDOA – FDOA geolocation has an interesting weakness in that for some amount of time per day, the two satellites used have very little differential frequency. This is due to the cyclical movement of the satellites. During those periods, FDOA measurements will not be ideal. In addition, the small amount of frequency difference being measured is much harder to accurately measure than the time differences.

TDOA-FDOA Geolocation

TDOA-FDOA geolocation, in most scenarios, gives ideal results. By combining time lines, which, generally, are oriented north-south, and frequency lines, which, generally, are orientated east-west, you get a nearly perpendicular crossing. A perpendicular crossing means less uncertainty in the calculated location. TDOA-FDOA geolocation also has an interesting limitation in that there are generally two times per day, separated by around 12 hours, where the FDOA becomes very small and hard to relate to an accurate LOP. These times can be calculated based on known satellite ephemeris information and approximate transmitter location, and can therefore be avoided when taking FDOA measurements.

The process of geolocating a signal requires some knowledge of the signal and all the techniques in order to get an accurate location.

The geolocation of a CW signal is nearly impossible with TDOA-FDOA. Nevertheless, a nominally CW transmission can contain imperfections, especially if a station transmits near its maximum EIRP. Hence, it often has a phase noise component which might be recognized as a modulated signal and therefore used to make TDOA measurements. However, it is generally more accurate to locate a CW carrier using FDOA-FDOA geolocation, even for non inclined satellites.

This is especially used today whenever high power CW jamming of actual full power multiplex transmissions occurs.

Syledis

Syledis (SYstem LEger pour mesure la DIStance) was a terrestrial radio navigation and locating system. The system operated in the UHF segment of 420-450 MHz. It was manufactured in France by Sercel S.A. and was operational during the 1980s and until about 1995, providing positioning and navigational support for the petroleum sector in the North Sea and to other scientific projects. Syledis has been replaced by GPS.

Functioning

Determination of the position of mobile vehicles, like f.e. vessels, using Syledis is accomplished by measurement of transit time of radio wavesbetween mobiles and radio stations at known points. There are two modes of operation, *active range mode* and *passive pseudo-range mode.*

Further Explanation of the Functioning of the Syledis Positioning System

A vessel is equipped with a transmitter that transmits a coded signal to at least three radio beacons each placed at a known point. The beacons send the code back to the transmitter. The returned coded signal is placed in a timeslot to determine the origin of the returned code. Therefore in an earlier stage a specified timeslot is connected to a specific beacon.

The elapsed time is proportional to twice the distance between the transmitter and the beacons. After the distance to the beacons is derived the position of the vessel can be calculated. The transmitter computes the distances to the beacons and a computer, connected with the transmitter, computes the position of the vessel.

Possible causes for Measurement Inaccuracy

The Syledis system has a measurement sensitivity that can be expressed in centimeters. Due to wheater conditions the wave propagation speed, for electromagnetic waves in air almost the speed of light, can change. In wet conditions, rain or snow, the wave propagation is a little bit slower than in dry conditions. So that gives an inaccuracy in determining the correct position of the vessel.

An other important factor is the length of the cables used to connect the antennas with the radio beacons at the shore, for example placed on lighthouses, or on oil rig platforms and the cables to connect the antenna with the transmitter at the vessel. Normally the antenna will be placed at the top of the mast and the transmitter wil be placed in the wheelhouse, where the captain and the mate can see the displays of the transmitter. Those cables has to be calibrated in order to obtain the smallest possible measurement uncertainty.

Real-time Geotagging

Real-time geotagging is a name given to the automatic technique of acquiring media (such as photos, audio or video), associating a specific location with the media, transferring the media to an online map and publishing the media in real time. It is thus an extension of an automatic geotagging process, requiring an in-built or attached location acquisitiondevice (such as GPS or Wi-Fi positioning system), but also requires communication with a wireless data transfer device (such as mobile phone networks or Wi-Fi networks). Several modern cell phones and digital cameras already integrate camera, aGPS, and wireless data transfer into one device,thus directly producing a geotagged photograph. Real-time geotagging is sometimes referred to as "mobile geotagging" or "autogeotagging", but this does not imply the real-time publishing step.

Location Acquisition

Geotagging is gaining popularity with photographers to produce geotagged photographs. A few cameras have built-in geolocation capability. Most, apart from smartphones, do not, so many photographers rely on external GPS receivers to determine location. Location may be inserted immediately into the picture file by tethering with Bluetooth or suitable wired connection, which are about as rare in cameras as the built-in autogeotagging feature.

GPS units, internal or external, require a lock from at least three satellites (for position), which usually requires up to 60 seconds. However, acquisition timeis decreasing rapidly with hardware improvements. Some 21st-century GPS receivers use cell tower location and one global positioning satellite to obtain a faster lock on location. This technology known as assisted GPS (aGPS) is becoming more popular in cellular phones, since it leverages cell tower locations.

One alternative to GPS is WiFi triangulation which uses the MAC addressesof nearby wireless access points to determine position. Automatic geotagging relies on media acquisition devices that contain GPS, aGPS, or WiFi or other local positioning systems. Many camera phones are also equipped to use the methods of mobile phone tracking.

Locations may be added later using a user's computer or a photo sharing web site such as Wikimedia Commons. It has even been proposed that in future, it may be possible to deduce the locations of photos purely by comparison with large numbers of geotagged photos. These geotagging techniques rely on post-processing of media, recorded tracks and/or MAC addresses, and cannot be used for real-time geotagging.

With the rapid rise of mobile smartphones which integrate GPS receiversand cameras, and relative market stagnation of separate devices, these phones are the most numerous instruments for autogeotagging. Those that don't include this ability in their stock software may acquire it by installing appropriate mobile apps.

Applications

Wireless data transfer allows real-time mapping of media. Transferring images from cell phones to social networking sites is gaining popularity. Applications for real-time mapping include travel, real estate, geosocial networking, people tracking, security, and geofencing.

Geotagging allows presenting the geotagged media via a map. Some examples of location-based social networking sites include MapWith.Us, Flickr, Panoramio and Picasa. However, presenting the data using online maps is a challenging problem, especially when combined with collaborative mapping. One example of such a project using real-time geotagging is MapWith.Us.

Privacy Concerns

Since real-time geotagging provides the real-time location of the person operating the device, it is possible to track that person using the data that they publish. Where this is a problem, the geotagger can choose to restrict online access to their data by means of access privileges.

Positioning Technology

Positioning systems will use positioning technology to determine the position and orientation of an object or person in a room, building or in the world.

Time of Flight

Time of flight systems determine the distance by measuring the time of propagation of pulsed signals between a transmitter and receiver. When distances of at least three locations are known, a fourth position can be determined using trilateration.

Optical trackers, such as laser ranging trackers suffer from line of sightproblems and their performance is adversely affected by ambient light and infrared radiation. On the other hand, they do not suffer from distortion effects in the presence of metals and can have high update rates because of the speed of light.

Ultrasonic trackers have a more limited range because of the loss of energy with the distance traveled. Also they are sensitive to ultrasonic ambient noise and have a low update rate. But the main advantage is that they do not need line of sight.

Systems using radio waves such as the Global navigation satellite system do not suffer ambient light, but still need line of sight.

Spatial Scan

A spatial scan system uses (optical) beacons and sensors. Two categories can be distinguished:

- Inside out systems where the beacon is placed at a fixed position in the environment and the sensor is on the object.

- Outside in systems where the beacons are on the target and the sensors are at a fixed position in the environment.

By aiming the sensor at the beacon the angle between them can be measured. With triangulation the position of the object can be determined.

Inertial Sensing

The main advantage of an inertial sensing is that it does not require an external reference. Instead it measures rotation with a gyroscope or position with an accelerometer with respect to a known starting position and orientation. Because these systems measure relative positions instead of absolute positions they can suffer from accumulated errors and therefore are subject to drift. A periodic re-calibration of the system will provide more accuracy.

Mechanical Linkage

This type of tracking system uses mechanical linkages between the reference and the target. Two types of linkages have been used. One is an assembly of mechanical parts that can each rotate, providing the user with multiple rotation capabilities. The orientation of the linkages is computed from the various linkage angles measured with incremental encoders or potentiometers. Other types of mechanical linkages are wires that are rolled in coils. A spring system ensures that the wires are tensed in order to measure the distance accurately. The degrees of freedom sensed by mechanical linkage trackers are dependent upon the constitution of the tracker's mechanical structure. While six degrees of freedom are most often provided, typically only a limited range of motions is possible because of the kinematics of the joints and the length of each link. Also, the weight and the deformation of the structure increase with the distance of the target from the reference and impose a limit on the working volume.

Phase Difference

Phase difference systems measure the shift in phase of an incoming signal from an emitter on a moving target compared to the phase of an incoming signal from a reference emitter. With this the relative motion of the emitter with respect to the receiver can be calculated Like inertial sensing systems, phase-difference systems can suffer from accumulated errors end therefore are subject to drift, but because the phase can be measured continuously they are able to generate high data rates.

Direct Field Sensing

Direct field sensing systems use a known field to derive orientation or position: A simple compass uses the Earth's magnetic field to know its orientation in two directions. An inclinometer uses the earth gravitational fieldto know its orientation in the remaining third direction. The field used for positioning does not need to originate from nature, however. A system of three electromagnets placed perpendicular to each other can define a spatial reference. On the receiver, three sensors measure the components of the field's flux received as a consequence of magnetic coupling. Based on these measures, the system determines the position and orientation of the receiver with respect to the emitters' reference.

Hybrid Systems

Because every technology has its pros and cons, most systems use more than one technology. A system based on relative position changes like the inertial system needs periodic calibration against a system with absolute position measurement. Systems combining two or more technologies are called hybrid positioning systems.

Vehicle Tracking System

A vehicle tracking system combines the use of automatic vehicle location in individual vehicles with software that collects these fleet data for a comprehensive picture of vehicle locations. Modern vehicle tracking systems commonly use GPS or GLONASS technology for locating the vehicle, but other types of automatic vehicle location technology can also be used. Vehicle information can be viewed on electronic maps via the Internet or specialized software. Urban public transit authorities are an increasingly common user of vehicle tracking systems, particularly in large cities.

Active versus Passive Tracking

Several types of vehicle tracking devices exist. Typically they are classified as "passive" and "active". "Passive" devices store GPS location, speed, heading and sometimes a trigger event such as key on/off, door open/closed. Once the vehicle returns to a predetermined point, the device is removed and the data downloaded to a computer for evaluation. Passive systems include auto download type that transfer data via wireless download. "Active" devices also collect the same information but usually transmit the data in near-real-time via cellular or satellite networks to a computer or data center for evaluation.

Many modern vehicle tracking devices combine both active and passive tracking abilities: when a cellular network is available and a tracking device is connected it transmits data to a server; when a network is not available the device stores data in internal memory and will transmit stored data to the server later when the network becomes available again.

Historically, vehicle tracking has been accomplished by installing a box into the vehicle, either self-powered with a battery or wired into the vehicle's power system. For detailed vehicle locating and tracking this is still the predominant method; however, many companies are increasingly interested in the emerging cell phone technologies that provide tracking of multiple entities, such as both a salesperson and their vehicle. These systems also offer tracking of calls, texts, web use and generally provide a wider range of options.

Typical Architecture

Major constituents of the GPS-based tracking are:

- GPS tracking: The device fits into the vehicle and captures the GPS location information apart from other vehicle information at regular intervals to a central server. Other vehicle information can include fuel amount, engine temperature, altitude, reverse geocoding, door open/close, tire pressure, cut off fuel, turn off ignition, turn on headlight, turn on taillight, battery status, GSM area code/cell code decoded, number of GPS satellites in view, glass open/close, fuel amount, emergency button status, cumulative idling, computed odometer, engine RPM, throttle position, GPRS status and a lot more. Capability of these devices actually decide the final capability of the whole tracking system; most vehicle tracking systems, in addition to providing the vehicle's location data, feature a wide range of communication ports that can be used to integrate other onboard systems, allowing to check their status and control or automate their operation.
- GPS tracking server: The tracking server has three responsibilities: receiving data from the GPS tracking unit, securely storing it, and serving this information on demand to the user.
- User interface: The UI determines how one will be able to access information, view vehicle data, and elicit important details from it.

Common Uses

Vehicle tracking systems are commonly used by fleet operators for fleet management functions such as fleet tracking, routing, dispatching, on-board information and security. Some vehicle tracking systems are bundled with or interface with fleet management software. Along with commercial fleet operators, urban transit agencies use the technology for a number of purposes, including monitoring schedule adherence of buses in service, triggering automatic changes of buses' destination sign displays once the vehicle approaches the bus terminus (or other set location along a bus route such as a particular bus stop along the route), and triggering pre-recorded (or even synthetic speech) bus stop, route (and its destination) or service announcements for passengers.

The American Public Transportation Association estimated that, at the beginning of 2009, around half of all transit buses in the United States were already using a GPS-based vehicle tracking system to trigger automated stop announcements. This can refer to external announcements (triggered by the opening of the bus's door) at a bus stop, announcing the vehicle's route number and destination, primarily for the benefit of visually impaired customers, or to internal announcements (to passengers already on board) identifying the next stop, as the bus (or tram) approaches a stop, or both; the latter are often also displayed on an internal LED display or LCD monitor connected to the system while the loudspeakers play them. Data collected as a transit vehicle follows its route is often continuously fed into a computer program which compares the vehicle's actual location and time with its schedule, and in turn produces a frequently updating display for the driver, telling him/her how early or late he/she is at any given time, potentially making it easier to adhere more closely to the published schedule.

Such programs are also used to provide customers with real-time information as to the waiting time until arrival of the next bus or tram/streetcar at a given stop, based on the nearest vehicles' actual progress at the time, rather than merely giving information as to the *scheduled* time of the next arrival.Transit systems providing this kind of information assign a unique number to each stop, and waiting passengers can obtain information by entering the stop number into an automated telephone system or an application on the transit system's website.

Some transit agencies provide a virtual map on their website, with icons depicting the current locations of buses in service on each route, for customers' information, while others provide such information only to dispatchers or other employees.

Other applications include monitoring driving behavior, such as an employer of an employee, or a parent with a teen driver.

Vehicle tracking systems are also popular in consumer vehicles as a theft prevention, monitoring and retrieval device. Police can simply follow the signal emitted by the tracking system and locate the stolen vehicle. When used as a security system, a Vehicle Tracking System may serve as either an addition to or replacement for a traditional car alarm. Some vehicle tracking systems make it possible to control the vehicle remotely, including block doors or engine in case of emergency. The existence of vehicle tracking device then can be used to reduce the insurance cost, because the loss-risk of the vehicle drops significantly.

Vehicle tracking systems are an integrated part of the "layered approach" to vehicle protection, recommended by the National Insurance Crime Bureau(NICB) to prevent motor vehicle theft. This approach recommends four layers of security based on the risk factors pertaining to a specific vehicle. Vehicle Tracking Systems are one such layer and are described by the NICB as "very effective" in helping police recover stolen vehicles.

Some vehicle tracking systems integrate several security systems, for example by sending an automatic alert to a phone or email if an alarm is triggered or the vehicle is moved without authorization, or when it leaves or enters a geofence.

Other scenarios in which this technology is employed include:

- Stolen vehicle recovery: Both consumer and commercial vehicles can be outfitted with RF or GPS units to allow police to do tracking and recovery. In the case of LoJack, the police can activate the tracking unit in the vehicle directly and follow tracking signals.
- Asset tracking: Companies needing to track valuable assets for insurance or other monitoring purposes can now plot the real-time asset location on a map and closely monitor movement and operating status.
- Field service management: Companies with a field service workforce for services such as repair or maintenance, must be able to plan field workers' time, schedule subsequent customer visits and be able to operate these departments efficiently. Vehicle tracking allows companies to quickly locate a field engineer and dispatch the closest one to meet a new customer request or provide site arrival information.
- Field sales: Mobile sales professionals can access real-time locations. For example, in unfamiliar areas, they can locate themselves as well as customers and prospects, get driving directions and add nearby last-minute appointments to itineraries. Benefits include increased productivity, reduced driving time and increased time spent with customers and prospects.
- Trailer tracking: Haulage and Logistics companies often operate lorries with detachable load carrying units. The part of the vehicle that drives the load is known as the cab and the load carrying unit is known as the trailer. There are different types of trailer used for different applications, e.g., flat bed, refrigerated, curtain sider, box container.
- Surveillance: A tracker may be placed on a vehicle to follow the vehicle's movements.
- Transit tracking: temporary tracking of assets or cargoes from one point to another. Users will ensure that the assets do not stop on route or do a U-Turn in order to ensure the security of the assets.
- Fuel Monitoring: monitor the fuel through tracking device (with help of fuel sensor connected to the device).
- Distance Calculation: calculate the distance travelled by the fleet.
- OBD II: Plug and play interface which provides most engine diagnostics information.

Vehicle tracking systems are widely used worldwide. Components come in various shapes and forms but most use GPS technology and GSM services. Additionally, these systems may also feature short range data communication systems such as WiFi. While most will offer real-time tracking, others record real time data and store it to be read, in a fashion similar to data loggers. Systems like these track and record and allow reports after certain points have been solved

Vehicle OBD Tracking Systems

Vehicle OBD tracking systems make use of OBD GPS trackers that plug into the on-board diagnostic (OBD) port of a light or medium duty vehicle. A cellular OBD GPS tracker directly communicates with the cell tower for sending the location and other vehicle performance data to the server over cellular wireless network. Usually, the tracker device draws power from the OBD port itself and contains a built-in antenna along with a GPS module for receiving the GPS signal. In addition, OBD trackers communicate with the different vehicle subsystems for receiving vehicle diagnostic and fuel consumption related data. Users can view the information using standalone software or web browser from a desktop/laptop computer or using smartphone apps.

Fleet Management Tracking

Aside from theft-prevention, the most common use of vehicle tracking is in logistics and transport. These systems make use of GPS and GSM technology to provide precise and constant location telematics to an individual fleet manager. These systems are typically equipped with features to monitor statistics such as; fuel consumption, average speed, current driver time and location. There has been a recent increase in demand for this technology as EU regulations place increased restrictions on the hours driver are allowed to work in a given day. It is currently limited to 9 hours per day. Companies are legally obligated to install a tachograph in any vehicle that is expected to carry goods. This obligation has led many to attempt to cauterize this potentially onerous obligation, instead turning it into a benefit. Fleet management systems use GPS & GSM technology. Much like other forms of trackers, although due to their nature they are equipped with more thorough diagnostic features.

Other uses such as Trailer Tracking, Fuel Monitoring, Distance Calculation, Asset Tracking, and Field Sales can all be incorporated into a Fleet Management solution.

Unconventional Uses

Industries not traditionally known to use vehicle tracking systems (logistics and transportation industries are the ones that have traditionally incorporated vehicle tracking system into their operations) have started to use it in creative ways to improve their processes or businesses.

The hospitality industry has caught on to this technology to improve customer service. For example, a luxury hotel in Singapore has installed vehicle tracking systems in their limousines to ensure they can welcome their VIPs when they reach the hotel.

Vehicle tracking systems used in food delivery vans may alert if the temperature of the refrigerated compartment moves outside of the range of safe food storage temperatures. Car rental companies are also using it to monitor their rental fleets.

W3C Geolocation API

The W3C Geolocation API is an effort by the World Wide Web Consortium(W3C) to standardize an interface to retrieve the geographical location information for a client-side device. It defines a set of objects, ECMAScriptstandard compliant, that executing in the client application give the client's device location through the consulting of Location Information Servers, which are transparent for the application programming interface (API). The most common sources of location information are IP address, Wi-Fi and BluetoothMAC address, radio-frequency identification (RFID), Wi-Fi connection location, or device Global Positioning System (GPS) and GSM/CDMA cell IDs. The location is returned with a given accuracy depending on the best location information source available.

Deployment in Web Browsers

Web pages can use the Geolocation API directly if the web browser implements it. Historically, some browsers could gain support via the Google Gears plugin, but this was discontinued in 2010 and the server-side API it depended on stopped responding in 2012.

The Geolocation API is ideally suited to web applications for mobile devices such as personal digital assistants (PDA) and smartphones. On desktop computers, the W3C Geolocation API works in Firefox since version 3.5, Google Chrome, Opera 10.6, Internet Explorer 9.0, and Safari 5. On mobile devices, it works on Android (firmware 2.0+), iOS, Windows Phoneand Maemo. The W3C Geolocation API is also supported by Opera Mobile 10.1 – available for Android and Symbian devices (S60 generations 3 & 5) since November 24, 2010.

Google Gears provided geolocation support for older and non-compliant browsers, including Internet Explorer 7.0+ as a Gears plugin, and Google Chrome which implemented Gears natively. It also supported geolocation on mobile devices as a plugin for the Android browser (pre version 2.0) and Opera Mobile for Windows Mobile. However, the Google Gears Geolocation API is incompatible with the W3C Geolocation API and is no longer supported.

Features

The result of W3C Geolocation API will usually give 4 location properties, including latitude and longitude (coordinates), altitude (height), and [accuracy of the position gathered], which all depend on the location sources. In some queries, altitude may yield or return no value.

Location Sources

The Geolocation API does not provide the location information. The location information is obtained by a device (such as a smartphone, PC or modem), which is then served by the API to be brought in browser. Usually geolocation will try to determine a device's position using one of these several methods.

Global Positioning System

This happens for any device which has GPS capabilities. A smartphone with GPS capabilities and set to high accuracy mode will be likely to obtain the location data from this. GPS calculate location information from the satellite signal. It has the highest accuracy; in most Android smartphones, the accuracy can be up to 10 metres.

Mobile Network Location

Mobile phone tracking is used if a cellphone or wireless modem is used without a GPS chip built in.

WiFi Positioning System

If WiFi is used indoors, a Wi-Fi positioning system is the likeliest source. Some WiFi spots have location services capabilities.

IP Address Location

Location is detected based on nearest Public IP Address on a device (which can be a computer, the router it is connected to, or the ISP the router uses). The location depends on the IP information available, but in many cases where the IP is hidden behind Internet Service Provider NAT, the accuracy is only to the level of a city, region or even country.

Implementation

Though the implementation is not specified, W3C Geolocation API is built on extant technologies, and is heavily influenced by Google Gears Geolocation API. Example: Firefox's Geolocation implementation uses Google's network location provider.

Google Gears Geolocation works by sending a set of parameters that could give a hint as to where the user's physical location is to a network location provider server, which is by default the one provided by Google (code.l.google.com). Some of the parameters are lists of sensed mobile cell towers and Wi-Fi networks, all with sensed signal strengths.

These parameters are encapsulated into a JavaScript Object Notation (JSON) message and sent to the network location provider via HTTP POST. Based on these parameters, the network location provider can calculate the location. Common uses for this location information include enforcing access controls, localizing and customizing content, analyzing traffic, contextual advertisingand preventing identity theft.

Example Code

Simple JavaScript code that checks if the browser has the Geolocation API implemented and then uses it to get the current position of the device. this code creates a function which can be called on HTML using <body onload="geoFindMe()">:

```
const geoFindMe = () => {
 if (navigator.geolocation) {
  navigator.geolocation.getCurrentPosition(success, error, geoOptions);
 } else {
  console.log("Geolocation services are not supported by your web browser".);
 }
}

const success = (position) => {
 const latitude = position.coords.latitude;
 const longitude = position.coords.longitude;
 const altitude = position.coords.altitude;
 const accuracy = position.coords.accuracy;
 console.log(`lat: ${latitude} long: ${longitude}`);
}

const error = (error) => {
 console.log(`Unable to retrieve your location due to ${error.code}: ${error.message}`);
}

const geoOptions = {
 enableHighAccuracy: true,
 maximumAge: 30000,
 timeout: 27000
};
```

Personal Trackers

The GNSS personal tracking is a product feature that allows persons to be tracked by a system that can make available the person's position to other systems or persons. This concept can be applied also to the tracking of objects or animals. The system works by attaching a GNSS device to the target to be tracked (e.g. person, object or animal). The GNSS device transmits positioning information to a server using independent communications means, usually the cellular network either by SMS or internet connection.

Personal trackers are an emerging trend, partially spread by the popular smartphones with built-in GNSS that can be used to locate a person with a relative lower cost.

Some of the most common motivation for individual tracking products includes monitoring children whereabouts, new born infants in hospital nurseries and elderly relatives. There are other tracking systems applications meant for crime control issues, pets tracking and wildlife tracking.

The electronic ankle bracelets are an example of personal tracking products, meant to control prison populations, house arrests or protecting victims of crimes.

The GNSS personal tracking devices transmit the location data and a unique id to a server using the GSM network allowing a large tracking coverage. These coordinates are used to determine the device location and present the location on a map service accessed from a website.

There are a number of reasons for GNSS tracking such as the location of individuals who get lost in outdoor activities, safety of children, elderly relatives or Parkinson patients.

Some devices give the possibility to monitor, not only the GNSS data (position, velocity and time) but they can also monitor heart rate, temperature or blood pressure, for people with health problems and special needs. Another feature available is the definition of a geofence, i.e., areas where the presence of the person is allowed, without generating an alert. In case the person trespasses outside the established zone, the device automatically begins tracking and sends an alert to a list of designated recipients usually via cell phone.

Some of these tracking equipment also own a "panic" button, allowing the tracked person to issue an alarm when in a dangerous situation such as being hurt, ill, lost or scared. When the alarm button is pushed in the device an alert is sent to a list of designated recipients. This information can also be transmitted to an emergency hotline.

Tracking systems have an enormous potential for fields other than children or elderly tracking, such as law enforcement, theft prevention, pets tracking or wildlife tracking, among others.

In terms of crime control issues, personal tracker products have potential use in the following situations:

- Law enforcement: This is the concept of keep tracking of a suspect, usually by attaching a GNSS device in a vehicle, in order to track the suspect's activities from a control room, preventing the police members to expose themselves and saving investigation resources.
- Theft prevention: This is the concept of keep tracking of companies' assets using the geofence mechanism. These systems raise an alert to the company's security or to the police once the tracked asset trespasses the control zone.

In terms of GNSS tracking in animals, there are the following common usages:

- Pets tracking: These are bracelets or collar systems targeted to owners of dogs and cats. The technology to use GNSS to track pets is relatively new. Most of current systems require utilization solely in areas covered by the GSM cellular network.
- Wildlife tracking: Meant for wildlife study and prevention, these tracking bracelets or collar systems are usually used in mammals and large birds.

The GNSS wildlife tracking is another high-potential growth area for GNSS applications. With ever smaller dimensions and weights and the availability of solar cell power these devices can be used in a growing number of cases. Simpler units store data in internal memory, which can be later retrieved.

Other systems for wildlife tracking units are able to send the data through a cell phone network in regular time intervals or on demand in the case of units with two way communication modes.

The most sophisticated GNSS wildlife tracking systems send the data in regular intervals via satellites, allowing a global coverage, almost in real time.

The applications for which personal trackers are used are considered as non-critical applications although in some cases can be used to ensure the safety of persons or goods.

The accuracy required by these applications is low. The current accuracy provided by civilian GPS is enough for these applications. Availability, however, can be reduced in urban environments or in areas with heavy foliage and deep canyons.

Product Characterization

The cost of these products has dropped considerably in recent years, offering the possibility of mass consumption by the average citizen. Aligned with this, the devices had demonstrated to be very useful in children monitoring. The tracking systems include a GNSS device to be placed in the tracking target.

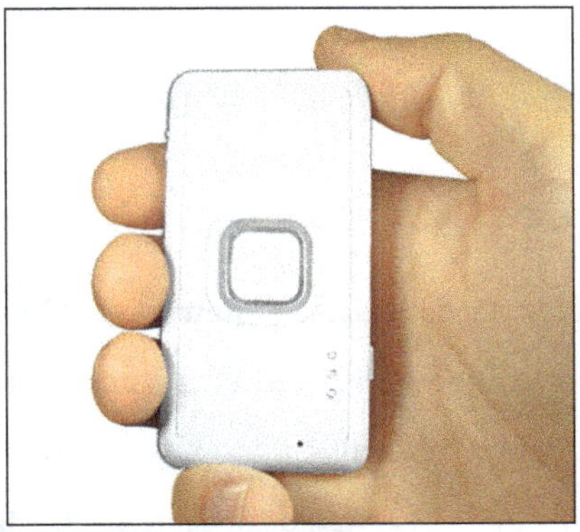

GNSS Tracker.

Usually, the tracking device is acquired as a service and is monitored by the location provider via cellular network, either using SMS or an internet connection, allowing the service to know always the person's location. This location is provided to the authorized persons through cell phone or through the provider's website.

Besides the cost of the device, usually there is a monthly subscription fee, similar to the way cellular phone service operates.

The Electronic Bracelets products are a specific tracking system used for the monitoring of prisoners, persons under house arrest or that have restrictions to their movement due to legal issues.

Detailed information about Electronic Bracelets can be found here.

Product Examples

The children tracking devices appear in several shapes and sizes, such as a wristwatch, a necklace, a cell phone and GNSS devices placed in book bags, shoes and clothing.

Usually the personal locaters are small standalone devices, but GNSS cell phones can be also used. Some cell phone operators provide the type of service either based on GNSS or in cellular network positioning.

Most personal trackers are sold in a package that might include the tracking services and that might require a subscription to a service. Normally these providers have a local nature having each country specific providers. There are too many providers to list here.

GNSS Wildlife tracking systems:

- Televilt: This company manufacture three different types of collars as well as backpacks for birds.

- Telemetry Solutions: A telemetry company involved in GNSS wildlife tracking since 1998.
- BlueSky Telemetry.
- Telonics.

References

- Geolocation, definition-1935: techopedia.com, Retrieved 31 March, 2020
- Automatic-vehicle-location, car-recovery: fleetistics.com, Retrieved 26 July, 2020
- Quick-start-indoor-positioning, solutions-basics: infsoft.com, Retrieved 17 August, 2020
- Dutton, Benjamin (2004). "15 – Basic Radio Navigation". Dutton's Nautical Navigation (15 ed.). Naval Institute Press. pp. 154–163. ISBN 155750248X
- Indoor-positioning-system: airfinder.com, Retrieved 13 April, 2020
- Personal-Trackers: gssc.esa.int, Retrieved 16 July, 2020
- Bolstad, Paul. GIS Fundamentals (PDF) (5th ed.). Atlas books. p. 102. ISBN 978-0-9717647-3-6

6
Applications

GNSS has a varied range of applications in transportation, precision agriculture, aerial photogrammetry, marine applications, geospatial information systems, rails, roads, etc. All these applications of GNSS have been carefully analyzed in this chapter.

Location based Services

Location-based services (LBS) use real-time geo-data from a mobile device or smartphone to provide information, entertainment or security. Some services allow consumers to "check in" at restaurants, coffee shops, stores, concerts, and other places or events. Often, businesses offer a reward – prizes, coupons or discounts – to people who check in. Google Maps, Foursquare, GetGlue, Yelp and Facebook Places are among the more popular services.

Location-based services use a smartphone's GPS technology to track a person's location, if that person has opted-in to allow the service to do that. After a smartphone user opts-in, the servic`e can identify his or her location down to a street address without the need for manual data entry.

Uses of Location-based Services

Companies have found several ways to use a device's location:

- Store locators: Using location-based intelligence, retail customers can quickly find the nearest store location.

- Proximity-based marketing: Local companies can push ads only to individuals within the same geographic location. Location-based mobile marketing delivers ads to potential customers within that city who might actually act on the information.
- Travel information: An LBS can deliver real-time information, such as traffic updates or weather reports, to the smartphone so the user can plan accordingly.
- Roadside assistance: In the event of a blown tire or accident, many roadside assistance companies provide an app that allows them to track your exact location without the need for giving directions.
- Mobile workforce management: For logistics-dependent companies that employ individuals out in the field or at multiple locations, LBS allows employees to check in at a location using their mobile device.
- Fraud prevention: An LBS creates another level of security by matching a customer's location through the smartphone to a credit card transaction. Tying the smartphone's location to a credit card allows you to flag transactions made across several geographic locations over a short time.

LBS Companies

Many app developers lack the resources to develop software to interpret a smartphone's location and instead use existing solutions via an API to save time and money. Many companies specialize in liaising with wireless carriers to connect companies with smartphone user locations. These companies provide tools to increase user engagement and connect with the most mobile phone users on the market. Companies well known for their LBS software include AT&T Mobile Marketing Solutions, Voxeo and Esri.

Personal Applications

The availability of low-cost GNSS chipsets allowed for integration of GNSS in different consumer products such as mobile phones. As a consequence, GNSS started being used not only for professional activities but also for leisure and entertainment purposes.

Some of these uses bring comfort to the user and can save the user time. Others are just to facilitate social interactions not bringing a tangible benefit to the user. The fact that the technology is available and free to use, allows for this kind of use.

Given the easy availability of GNSS positioning in consumer products the use of GNSS positioning for personal applications has become usual and new uses for the technology are constantly appearing.

Pedestrian Navigation

Pedestrian Navigation consists in the navigation of pedestrians in urban environments. Pedestrian navigation often offers pedestrian specific maps and multi-modal navigation that combines walking directions with all available modes of transportation and even routing in indoor environments.

Pedestrian navigation requires more complex maps and algorithms to provide an adequate navigation experience than road navigation which is a simpler, much more predictable and contained environment.

Outdoor Navigation

Outdoor Navigation consists in the navigation in outdoor environments such as nature trails. Outdoor Navigation is used mainly by outdoor enthusiasts (such as hikers, backpackers, geocaching enthusiasts, mountaineers or bikers) and outdoor workers (such emergency workers, wildlife preservation workers, forestry workers, farmers or mining workers).

Outdoor Navigation is somewhat different from Pedestrian Navigation or Road Navigation since it uses the concepts of trails and waypoints instead of a turn by turn navigation. Maps are not always required since individual trails can be loaded on the device or trails can be created from hikes. When present, maps are very different from the ones used in Pedestrian Navigation or Road Navigation; outdoor navigation maps more closely related with topographic charts and military charts, rather than being street and road oriented.

Social Networking

Social Networking has become a part of our daily life and millions of cybernauts use social network to interact with their social and professional contacts. The concept of posting information about "what are you doing" very rapidly progressed to post "where are you".

Location-enabled Social Networking allows users to post their current location or to check into places or venues. This allows the user to disseminate his/her location to his/her friends while receiving the information of who is in the same vicinities.

Photography Geocoding

The popularity of applications, such as Google Earth and Google Maps, and the possibility of posting georeferenced pictures lead to the need of geocoding photos. A geocoded photo has the location of the photographer embedded in the files. This information includes at least latitude and longitude but might include also altitude and the bearing of the camera.

The most practical way of geocoding photos is by doing it automatically on the camera if the camera has an embedded GNSS receiver or can be connected to an add-on GNSS receiver. GNSS-enabled mobile phones usually geotag photos automatically.

Location based Services

Location Based Services (LBS) include applications that depend on the user location to provide a service/information that is relevant to the user at that location.

Road Applications

The road sector is a major potential market for GNSS applications. Satellite navigation receivers are now commonly installed in new cars as a key tool for providing new services to people on the move: electronic charging, real-time traffic information, emergency calls, route guidance, fleet management and Advanced Driving Assistance Systems.

Satellite navigation will help regulate road use and minimize traffic jams. If all vehicles are fitted with a navigation satellite receiver and a data transmitter, their position can be relayed automatically every few seconds to a central station. This information can then be used in a number of ways to control road usage. It could, for example, be used to charge motorists for using a stretch of road, to restrict access to congested roads, or to inform drivers of congestion and suggest alternative, quieter routes.

The road sector is a major potential market for GNSS applications. Satellite navigation receivers are now commonly installed in new cars as a key tool for providing new services to people on the move: electronic charging, real-time traffic information, emergency calls, route guidance, fleet management and Advanced Driving Assistance Systems.

Satellite navigation will help regulate road use and minimize traffic jams. If all vehicles are fitted with a navigation satellite receiver and a data transmitter, their position can be relayed automatically every few seconds to a central station. This information can then be used in a number of ways to control road usage. It could, for example, be used to charge motorists for using a stretch of road, to restrict access to congested roads, or to inform drivers of congestion and suggest alternative, quieter routes.

Road Navigation

Car navigation is still one of the main applications of GNSS, as reported in the GNSS Market Report, Issue 3. GNSS data combined with electronic maps allow positioning and guidance of the road user.

Route guidance using satellite navigation is already a well-established product offered both by car manufacturers and standalone navigation devices. The majority of these systems are based on satellite navigation systems that can be integrated with onboard sensors (odometer and gyros) to compute optimal routes in real-time.

Road Navigation.

Tolling

It is foreseen that traditional ways of road tolling will be substituted by solutions based on positioning by means of GNSS. On the other hand there's the need of implement urban tolling that is usually called Congestion Charge Zone (CCZ) with the aim to avoid urban pollution and urban traffic congestion. The use of GNSS for Electronic Tolling Collection (ETC) has many advantages with respect to other technologies as it allows for a flexible and scalable system with minimum specific roadside infrastructure. This translates directly into a simple and cost efficient system. In fact, the interoperability directive 2004/52/EC adopted by EU in 2004 prescribed the development and deployment of a European Electronic Tolling Service (EETS), recommending the GNSS technology as the one to be adopted.

Road and urban tolling based on GNSS implies that the position and trajectory of a vehicle is determined using GNSS in order to decide if the vehicle must be charged or not and to compute the charging value. This must be done along with the determination of an accuracy level and the acquisition/generation of some kind of irrefutable evidences that can prove the amount of applied charges.

Emergency Services

In automobile crashes, one of the most important factors in saving lives is a prompt response from emergency medical services. By improving information transfer between the trauma care physician and emergency medical service personnel they result in faster, more appropriate care, thus saving lives and reducing disabilities resulting from crashes. Traffic incident and emergency management is the planned and coordinated program process to detect, respond to and remove traffic incidents and restore traffic capacity as safely and quickly as possible.

The use of GNSS for emergency services and incident management can make the response to emergency situations much faster and efficient saving lives. The precise location of vehicle can be sent to rescue authorities and can use the emergency and rescue vehicle fleet management system to assign the most adequate vehicle to respond to the incident.

Emergency Services.

Traffic Management

Traffic Management.

Congestion of public road networks is a growing problem in many countries. Authorities are developing initiatives to manage the traffic, but no remedial strategy can be better than the information upon which it has to rely. Consequently, the traffic planners need information that is accurate, reliable, timely and complete. The road users too need good-quality traffic information in order to plan and adjust their routes.

The monitoring and management of traffic fluidity will be significantly facilitated when a great number of cars are equipped with satellite navigation receivers and guidance systems. For example, if the average speed of the cars equipped with GNSS receivers in a road sector drops significantly, a control center can anticipate a traffic jam and suggest that approaching vehicles choose different routes. Several studies have concluded that travel time would be cut by 10-20%.

Fleet Management and Vehicle Tracking

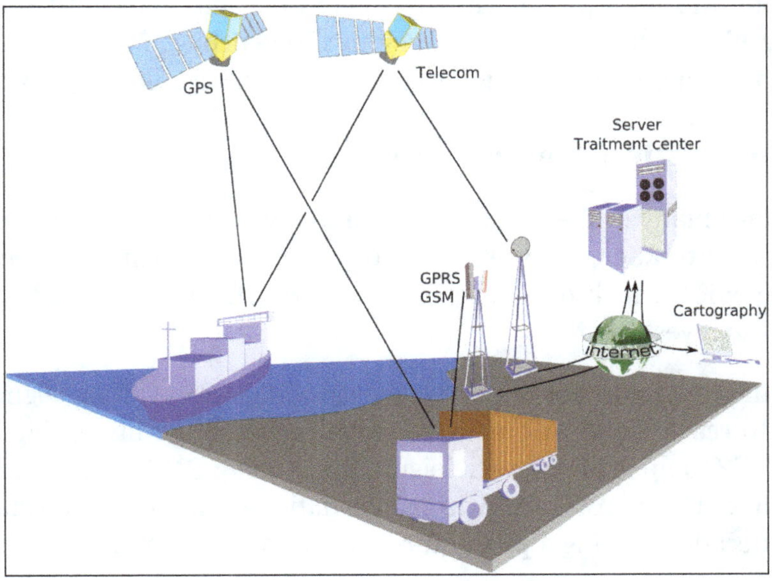

Fleet management and vehicle tracking.

GNSS is being used or the tracking and tracing of goods, vehicle scheduling and control and improved "just-in-time" delivery processes. Modern economies performance is very closely linked to the performance of their freight and fleet management systems. The free movement of goods relies heavily on road transport and the efficiency and effectiveness of road-based distribution has a major impact on transport systems.

GNSS-based fleet management systems are used to locate vehicles (e.g. trucks, buses, police cars, taxis) in order to optimize resource management, reduce travel time, increase security and reduce fuel consumption. Usually these systems are based on embedded devices and On-Board Units (OBUs) that are continuously growing and replacing the nomadic PND. As a consequence of this significant growing new Intelligent Transport System (ITS) services are expected to be deployed in the coming years, taking the use of GNSS far beyond in-vehicle navigation.

Aviation Applications

In aeronautical applications, satellite navigation has long been an additional means for localization purposes. The development of GNSS has provided a supplementary positioning service for many flight phases, in leisure flying as well as commercial air transport. Refining and improving satellite navigation through augmentation systems will assist pilots in all flight phases, from taxing, to take-off, en-route flying, and landing in all weather conditions, reaching the level of safety that will be required to cope with the continuous increase in the number of flights.

When planning aircraft routes and landing schedules at busy airports, it is essential to ensure that aircrafts are always a safe distance from each other. GNSS, will provide a guaranteed service with sufficient accuracy to allow airlines and pilots to know their position reliably and precisely enough to make substantial efficiency savings.

Overview of Aviation Applications

GNSS overcomes many of the inadequacies in today's air traffic infrastructure thanks to its accurate, continuous, all-weather positioning service. This is especially relevant for General Aviation which often uses smaller airports, where the adequate systems represent a large investment.

According to the GNSS Market Report, Issue 3, total GNSS aviation segment revenues are expected to reach €900 mln in 2022 driven by revenue increasing of the Visual Flight Rules (VFR) sub-segment, with leisure flyers using GNSS as a supplementary information source. The use of GNSS within all aviation segments is expected to increase over the next decade reaching a penetration of over 90% in 2022.

The original basic GNSS equipment is evolving due to the development of augmentation systems such as SBAS, GBAS and ABAS, while the introduction of Galileo and the modernization of GPS and GLONASS will further improve GNSS performance. The use of systems that integrate GNSS and inertial systems is also expanding.

The following section provides an overview of augmentation systems used in aviation applications.

GNSS Augmentation

Aviation GNSS applications and products have been one of the main drivers of GNSS Augmentation systems. The ICAO GNSS Standards defines three architectures to augment GNSS constellations for aviation, ABAS (Aircraft Based Augmentation Systems), GBAS (Ground Based Augmentation System) and SBAS (Satellite Based Augmentation System).

Present ground-based navigation aids are monitored, and the monitor takes action if erroneous signals are being radiated. On the present configuration of the GNSS system it may take considerable time before users become aware of any malfunctioning.

ABAS

The Aircraft based augmentation can provide GNSS information as necessary for supplemental means of navigation. An ABAS is basically a system that augments and/or integrates the information obtained from the other GNSS elements with information available on board the aircraft.

The Aircraft based augmentation system can be implemented by:

- Receiver Autonomous Integrity Monitoring (RAIM) whereby a GNSS receiver processor determines the integrity of GNSS position using GPS signals or GPS signals augmented with altitude (baroaiding).
- Aircraft Autonomous Integrity Monitoring (AAIM) whereby the GNSS signal is integrated with other navigation sensors, such as the inertial navigation system (INS) which can perform an integrity check on GNSS data when RAIM is unavailable.

  Nowadays, around 70% of European flights are made by aircraft equipped with GPS and RAIM.

GBAS

The GBAS provide GNSS integrity monitoring through data obtained from the ground. They also increase the accuracy of satellite navigation, clearing the way for GNSS precision approach and landing. A ground station at the airport transmits locally-relevant corrections, integrity data and approach data to aircraft at the terminal area in the VHF band. GBAS will increasingly support CAT II/III operations where economically beneficial once enhanced GNSS become available. It is assumed that all GBAS stations for CAT I operations will be upgraded to CATII/III operations.

SBAS

SBAS systems implemented in in several regions improves the accuracy, reliability and integrity of the GPS signal. GNSS-SBAS navigators that meet international organizations regulations (such as ICAO, FAA, Euro control) may be used for sole means of navigation for all phases of flight, including precision approach at airports.

The SBAS systems offer an opportunity for airports to gain Instrument Landing System (ILS) approach capability without the purchase or installation of any ground-based navigation equipment at the airport.

The SBAS vertically guided approach procedures are considered LPV (Localizer Performance with Vertical guidance) and provide ILS (Instrument Landing System) equivalent approach minimums as low as 200 feet at qualifying airports. Actual minimums are based on an airport's current infrastructure, as well as an evaluation of any existing obstructions.

In Europe the EGNOS Safety-of-Life signal was formally declared available to aviation in March 2011. For the first time, space-based navigation signals augmented by this system have become officially usable for the critical task of vertically guiding aircraft during landing approaches.

En-Route Navigation

GNSS overcomes many of the deficiencies in today's air traffic infrastructure thanks to its accurate, continuous, all-weather positioning. During en-route flight, the availability of GNSS will ensure high robustness through the redundancy and high reliability of the service.

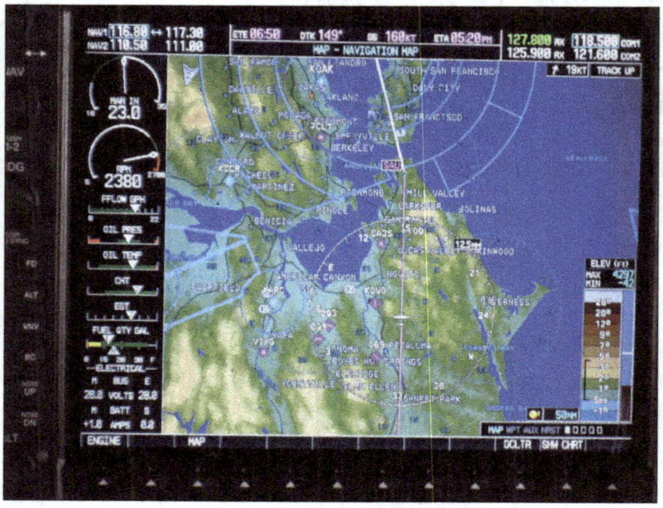

En Route Navigation.

In the future, higher accuracy and service integrity will allow aircraft separation to be reduced in congested airspace, to cope with traffic growth. GNSS will be used in en-route flight phase of commercial aircraft.

Approach

Portable GNSS device.

Approach, landing and take-off are critical flight phases and the major need of commercial operators is to have full operations in all weather conditions. As a consequence, precision approach is a mandatory requirement.

The GNSS, with the aid of ground-based augmentation (local elements), will satisfy the needs for precision approach as defined in the aeronautical standards, and could replace or complement the navigation infrastructure of airports in regions where the system is inadequate. The most prominent example is the airports that are not equipped with instrument landing system.

Attitude Determination

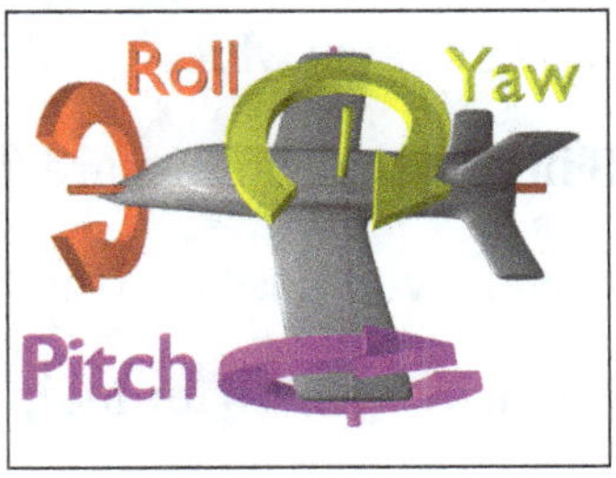

Flight dynamics.

The Attitude Determination is one of the many applications where GNSS can be effectively employed.

The attitude of an aircraft, i.e., the orientation in space, can be determined by measuring the relative positions of multiple GNSS antennas mounted on different positions of the aircraft. Usually, a set of 3 or more GNSS antennas placed on board of an aircraft can provide the complete information to compute the aircraft's attitude.

Air Traffic Control

Air traffic controllers need position, heading, speed and time information for the continuous management of all aircraft. Some areas of the world lack the appropriate ground infrastructure, including secondary radar and communication links. Standardized transmission of GNSS navigation data will lead to advanced systems and techniques for safer air traffic monitoring.

The Automatic dependent surveillance-broadcast (ADS-B) is currently the most important system used in ATC side, which relies on GNSS as primary data source, to obtain aircraft's horizontal positions.

Rail Applications

Rail systems throughout the world are beginning to use GNSS to track the movement of locomotives, rail cars, maintenance vehicles, and wayside equipment in real time. When combined with other sensors, computers, and communications systems, GNSS improves rail safety, security, and operational effectiveness. The technology helps reduce accidents, delays and operating costs while increasing track capacity, customer satisfaction and cost effectiveness.

The railway domain could considerably profit from the implementation of autonomous on-board positioning systems. Especially for local and regional railway lines, there are various possible applications which could enable a cost-effective modernization and increase of efficiency. However, train control poses high demands on positioning with respect to availability, reliability and integrity. These requirements can only be fulfilled by means of integrated positioning systems which combine GNSS with other sensors.

The benefits of the use of GNSS in rail applications are the following:

- The increasing of situational awareness for train's safety.
- The increasing of situational awareness for maintenance crews.
- The prevention of collisions, derailments, work zone incursions, and rail switch errors.
- The increasing of capacity and efficiency for all rail users.
- Dependable schedule and equipment location awareness.
- Automated track surveys and inspections.
- The time synchronization for communication systems.

In the railway sector, usually non-safety and safety-critical applications are distinguished. Safety-critical applications which include train control and signalling, passenger traffic or transportation of dangerous goods, show very demanding requirements in terms of availability, continuity and integrity.

Safety-critical Applications

Safety-critical applications include e.g. train control and signaling, which in Europe have been subjects to a standardization process in recent years in order to improve interoperability.

In particular, European standard for train control, signaling and traffic management called the European Rail Traffic Management System (ERTMS) is emerging. This standard includes two layers that can rely on satellite navigation: the European Train Control System (ETCS) and the European Traffic Management Layer (ETML).

Other safety-critical examples where accurate positioning data provided by GNSS can be used are the rail track surveying, which enable real-time performance monitoring contribute significantly to safety since they allow quicker detection of defects than traditional methods which are based on separate periodical track inspections.

Finally, the use of GNSS data in assisting car body tilt control, allows tilting trains to negotiate sharp curves at higher speed (thereby enabling shorter traveling times) and with less centrifugal force inside a car. Detecting the accurate running position is essential here and traditionally it has been calculated by monitoring the number of wheel revolutions and using the ground coils installed on the track as position reference points. However, GNSS-based systems have been recently developed which allow a positioning accuracy equal to or higher than the conventional system.

Non-safety-critical Applications

Non-safety-critical applications, in turn, are mainly related to fleet management, goods tracking and other logistic information management. The ability to effectively track the location of goods and to estimate their delivery times is indeed as important in railway sector as in other modes of transportation. On the other hand, GNSS-based fleet management will also help to improve the performance of passenger transportation. Providing up-to-date information about the arrival and departure times of trains, especially when there are delays, is an essential part of good service.

Traffic Management and Signaling

Train control systems pose high demands on positioning with respect to availability, reliability and integrity. These requirements can only be fulfilled by means of integrated positioning systems, which combine GNSS with other sensors.

The use of GNSS in railway systems presents many advantages, in particular the monitoring of the train exact location, logistic information management, enhanced train signaling (which improves safety, but also enables e.g. reduced distances between trains and therefore increased train frequencies), and the possibility to map the transport infrastructure.

Thus, while the number of applications based on GNSS is considerably behind the number of those used in other domains, such as road transport, incorporating GNSS receivers into modern signaling, train control and other railway systems has become common.

Maritime Applications

According with International Maritime Organization (IMO), *Shipping is the most international of all the world's great industries*. In fact a huge number of vessels move around the world each day, making the efficiency, safety and optimization of marine transportation key issues.

GNSS technology will be fundamental for bringing innovation and progress in navigation and many other marine activities such as fishing, oceanography and oil and gas exploitation. Satellite navigation benefits all maritime applications, including leisure boats, commercial vessels, and unregulated and Safety of Life at Sea (SOLAS) regulated ships. Each application will take advantage of the new characteristics offered by GNSS augmentation: increased accuracy and integrity, certified services and high availability.

GNSS will be used in every phase of marine navigation: ocean, coastal, port approach and port manoeuvres, under all weather conditions.

GNSS is used mainly in maritime en route navigation and in support to the AIS and VTS systems which are maritime traffic collision and traffic control mechanisms. In maritime en route navigation there are two categories of applications, the chart plotters receivers and fish finder applications.

Besides the above mentioned applications, there are several other utilizations of GNSS in maritime applications, such as:

- Observing the changes of sea level,
- Dredging operations,
- Wrecks location,
- Laying pipe lines,
- Search and Rescue of sinking vessels,
- Dynamic positioning,
- Positioning of oil rigs,
- Fixing of satellite sea launch platforms.

En-Route Navigation

Nowadays, ship navigation casualties and incidents can result in serious loss of life and pollution of the marine environment as modern ship can carry over 5,000 people and over 500,000 tons of petroleum. A wide variety of vessels moves around the world each day. The efficiency, safety and optimization of marine transportation are key issues.

Due to the presented reasons, GNSS is being adopted as a marine navigation tool because it provides a reliable and short cost solution, allowing interoperability across navigational areas and with great impact in safety and commercial benefits, for all maritime purposes, including leisure boats, commercial vessels, unregulated and Safety of Life at Sea (SOLAS) regulated ships.

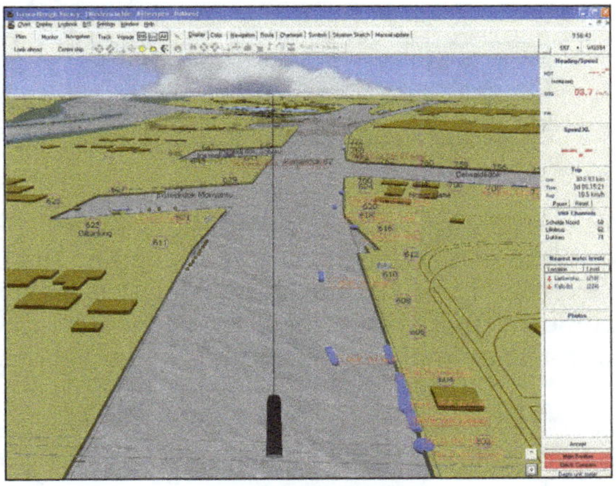

En route navigation.

AIS/VTS

Vessel traffic system.

The Automatic Identification System (AIS) is perhaps the biggest technological advance in maritime navigation, since the introduction of radar. The service is a shipboard broadcast system that acts like a transponder, operating in the VHF maritime band and transmitting real time information of the vessel, in a similar manner as ADS-B for aviation. In the same way as ADS-B, the AIS rely on GNSS as primary positioning source.

The Vessel Traffic Service (VTS) provides active monitoring and navigational advice for vessels in particularly confined and busy waterways. Usually the VTS integrates in a command and control room, information from sources such as radar, AIS and closed circuit television sites.

Permissions

All chapters in this book are published with permission under the Creative Commor Attribution Share Alike License or equivalent. Every chapter published in this book ha been scrutinized by our experts. Their significance has been extensively debated. The topic covered herein carry significant information for a comprehensive understanding. They ma even be implemented as practical applications or may be referred to as a beginning poir for further studies.

We would like to thank the editorial team for lending their expertise to make the bool truly unique. They have played a crucial role in the development of this book. Without thei invaluable contributions this book wouldn't have been possible. They have made vital efforts to compile up to date information on the varied aspects of this subject to make this book a valuable addition to the collection of many professionals and students.

This book was conceptualized with the vision of imparting up-to-date and integrated information in this field. To ensure the same, a matchless editorial board was set up. Every individual on the board went through rigorous rounds of assessment to prove their worth. After which they invested a large part of their time researching and compiling the most relevant data for our readers.

The editorial board has been involved in producing this book since its inception. They have spent rigorous hours researching and exploring the diverse topics which have resulted in the successful publishing of this book. They have passed on their knowledge of decades through this book. To expedite this challenging task, the publisher supported the team at every step. A small team of assistant editors was also appointed to further simplify the editing procedure and attain best results for the readers.

Apart from the editorial board, the designing team has also invested a significant amount of their time in understanding the subject and creating the most relevant covers. They scrutinized every image to scout for the most suitable representation of the subject and create an appropriate cover for the book.

The publishing team has been an ardent support to the editorial, designing and production team. Their endless efforts to recruit the best for this project, has resulted in the accomplishment of this book. They are a veteran in the field of academics and their pool of knowledge is as vast as their experience in printing. Their expertise and guidance has proved useful at every step. Their uncompromising quality standards have made this book an exceptional effort. Their encouragement from time to time has been an inspiration for everyone.

The publisher and the editorial board hope that this book will prove to be a valuable piece of knowledge for students, practitioners and scholars across the globe.

Index